We would like to extend our gratitude to everyone who contributed to the Institute's 75th Anniversary Book Project through oral history interviews, donations of photographs and other memorabilia, monetary gifts, and general good wishes.

Author, Lead Historian
Megan Murray Wilhoit

Production Editor
Cindy Warden Bowden
Megan Sheppard

Research Historian
Bonnie Sue Boyd

Editor
Peter Hopkins

Graphic Designers
Gina Fontana-Siepierski
Julie Sanders

Research
David Hamilton
Qingfen Pan

Photography
David Bell
Tom Bowden
Robert Hill
Kara Kynast
Karl Warden

Printing
Printed in China by
Sun Fung Offset Binding Co. Ltd.

Publisher
Sun Fung Museum Catalogs & Books,
Sarasota

Production
Connie Adams
George Atkins
Charles Andry
David Bell
Christina Bowden
Charles Brookshire
Juan Chevere
Bob Davies
Perry Dear
Owen Dundee
Lavon Harper
Farnaz Jafari
Patrick Johnson
Jessica Kirk
Megan Kirk
David Pena
Fran Rottenberg
Santiago Santamarina
Laurie Sedicino
Megan Sheppard
Doug Stone
Matt Tennihan
Teresa Williams
Robin Young

SUN FUNG
Museum Catalogs & Books

Table of Contents

Credits and Acknowledgements i

Table of Contents ii

Book Sponsorships iv

Forward: Dr. William J. Frederick, Jr. vi

Introduction: 1920-1929 2

Chapter 1: 1930-1939 18

Chapter 2: 1940-1949 38

Chapter 3: 1950-1959 58

Chapter 4: 1960-1969 78

Chapter 5: 1970-1979 96

Chapter 6: 1980-1989 110

Chapter 7: 1990-1999 124

Chapter 8: 2000-present 138

The Robert C. Williams Paper Museum at Georgia Tech 148

BOOK SPONSORSHIPS

Legacy Donors

George W. Mead
IPC Foundation
IPST Alumni Association
Georgia-Pacific Corporation

Platinum Level

William A. Finn
J. Carter Fox
Woody Rice
Sondra and Charles Gilman, Jr. Foundation, Inc.

Gold Level

William R. Haselton
Robert W. Hisey
David G. McMaster
Richard E. Olson
Richard W. Roberts
William H. Trice

Silver Level

A. Nurhan Becidyan
Steven L. Edwards
David Fary, Jr.
William J. Frederick, Jr.
Arthur J. "Ben"Haug
Carl A. Jentzen
Joseph V. Koleske
Aarre Metsavirta
W. Thomas Myers, Jr.
Paul Saueracker
John M. Vaughan

Bronze Level

Peter Balousek
Richard G. Barker
John S. Barton
C. Nevin Betts
Robert L. Boehm
Keith E. Bradway
Alan Button
Jeffrey and Gail Carels
Jacinta Castellino
Thomas E. Crozier
Thomas P. Czepiel
Roy L. Davis
Arthur Dreshfield
Richard C. Erickson
James C. Farrand
James D. Freiberg
Margaret S. Gander
David P. Graves, Jr.
Brenton S. Halsey
Gerald R. Haw
Thomas W. Johnson
Robert L. Jones
Daniel Kane
Stuart E. Lebo
Ismo Lepola
James E. Macklem
George Martin
Victor F. Mattson
John M. McEwen
Timothy McIlvenna
David McIntyre
Jer Fei Mih
John Parkinson
John Rogers
Vincent Russo
SAPPI Fine Paper
S.D.Warren Company
Harry Seamans
Harry L. Spiegelberg
James S. Street
Donald W. Theisen
Nick Triantafillopoulos
Karen Turnboom
Larry W.Wanamaker
William H.Wiseman

Dear Institute Family And Friends,

We are pleased and proud to present to you *From Appleton to Atlanta: The Institute's First Seventy-Five Years*. This product of labor and love is the result of hundreds of hours of oral history interviews with former and current Institute students, faculty and staff members, administrators, and friends as well as a myriad of Institute memorabilia from the museum's archives.

This project would not have been possible without the time and effort of many interested and invested parties. Megan Murray Wilhoit, the book's author and lead historian, has worked on this project for more than three years and has done a wonderful job transforming interviews, photographs, and a wide-ranging series of documents pertaining to the Institute's history into what you now hold in your hands. Cindy Bowden, the director of the Robert C. Williams Paper Museum at Georgia Tech, has dedicated endless hours to the book's production, campaigning tirelessly for the support needed to complete the project. To Megan and Cindy, as well as to the rest of the museum staff and others at the Institute who have supported this project, we extend our deepest gratitude.

I would also like to thank our Legacy Donors - George W. Mead, the IPC Foundation, the IPST Alumni Association, and the Georgia-Pacific Corporation – as well as all of the other Institute alumni and friends whose financial contributions have made this project possible. Your support will ensure that the Institute's rich and proud legacy will be forever remembered.

Sincerely,

DR. WILLIAM J. FREDERICK, JR.
Director
Institute of Paper Science and Technology at Georgia Tech

INTRODUCTION

1920-1929

The 1920s was a decade renowned for gaiety and flamboyance, an era of unparalleled prosperity and rapid change. American women, emboldened by their newfound right to vote, embraced the fashion of the "Flapper" and could be found dancing the Charleston with young World War I veterans as jazz tumbled out of nightclubs and speakeasies across the country. Prohibition, brought on by the 1919 Volstead Act and codified by the Eighteenth Amendment, ushered in the legendary and oft-romanticized era of gangsters making substantial livings controlling the flow of illegal liquor in cities across the country. American consumerism reached a fevered pitch and manufacturers raced to keep pace with the ever-increasing demand for automobiles, telephones, refrigerators, and radios.

And then it was over.

Americans crowded around their newly-purchased radios and listened, stunned, as news of a stock market crash, the worst in the nation's history, raced across the airwaves. That fateful day, "Black Tuesday," October 29, 1929, marked the end of the lavish splendor of the Roaring Twenties and ushered in one of the most challenging eras in United States history. The Great Depression, which cost some thirteen million jobs within the first two years, wreaked havoc on the American economy. Businesses folded at an alarming rate and plans for new ventures were abandoned for lack of capital.

Except for one.

Milwaukee was the site of Wisconsin's first paper mill. On March 7, 1848, three months before Wisconsin was officially granted statehood, papermakers at a mill located at the junction of the Milwaukee and Menomonee Rivers used recycled cotton rags to produce paper for the *Milwaukee Sentinel and Gazette* newspaper. Production eventually climbed to ninety reams per week.

BUT SOUTHERN WISCONSIN was not to become the center of the state's papermaking industry. Papermakers, following the bountiful supply of fresh water from the Fox, Wisconsin, Chippewa, Menominee, Peshtigo, Eau Claire, and Flambeau Rivers, began to move northward, transforming abandoned flour and lumber mills into paper mills.

In 1853, the Richmond brothers established the Fox River Valley's first paper mill in the town of Appleton. Its primary products were a coarse-grain rag paper and wrapping paper made from straw. The mill, like so many before and so many to follow, burned in 1859, but was rebuilt at once and began producing more than four tons of wrapper and manila paper per day. Within the next twenty-five years, additional mills were built at Kaukauna, Neenah, and Fond du Lac, establishing the Valley's reputation as the papermaking center of Wisconsin.

1872 proved a turning point for what was quickly becoming Wisconsin's chief industry. In the latter part of that year, Col. H.A. Frambach, a Civil War veteran, introduced to the state the Keller Groundwood Process. Developed in Germany in 1844 by Friedrich Keller, the process involved pressing logs against a grinder and adding water to create suitable fibers for papermaking, heralding the end of constant cotton rag shortages. The beautiful, untouched North Woods promised enough lumber to see the Wisconsin paper industry well into the next century. The Frambach and Stovekin Mill in Kaukauna, founded in 1873, was the first to install a machine for grinding pulpwood logs and others quickly followed suit.

Westbrook Steele

Westbrook Steele arrived in Appleton in 1928 to assist Lawrence College in its ambitious reorganization program. He left twenty-seven years later as President Emeritus of the Institute of Paper Chemistry, the graduate school to which he had contributed enormous amounts of time, energy, and plain, old-fashioned sweat.

Steele was no stranger to educational experiments. He had spent more than twenty years as general manager of National Systems, a company that directed the reorganization of schools and colleges. His record boasted the successful restructuring of more than seventy educational institutions.

Steele's contributions to the Institute are legendary. It was he who persuaded local paper manufacturers to donate equipment for the original pulping laboratory. He convinced allied industries to furnish the money needed to complete the Institute's first building, and, in one of his final acts as the Institute's President, arranged the acquisition of the Dard Hunter Collection of paper and paper-related artifacts. Institute alumni will forever remember the welcoming tea party held each September in the Steeles' home and the cheerful holiday parties for all the Institute children hosted by Mrs. Steele each Christmas.

During Steele's retirement celebration in September of 1955, Dean Harry Lewis, a long-time colleague and friend, aptly summarized Steele's Institute of Paper Chemistry experience:

His dreams have been realized in a very real way.

The Keller Groundwood Process connected Wisconsin's thriving paper industry with its other premier industry. Logging was slow to develop in Wisconsin, with the area's earliest white settlers preferring to bring lumber with them from the East. After 1836, however, when the Menominee Indians were forced to cede much of the land in central and eastern Wisconsin to the United States, the state's logging industry began to develop in earnest. Like the paper industry, logging was directly dependent on the state's numerous rivers and waterways, which provided convenient transport for pine logs from the wild forests of central and northern Wisconsin to the mills farther south. Dozens of new paper mills sprang up in central and northern Wisconsin in the years leading up to the dawning of the 20th century, including several still in operation today.

Even into the new century, papermaking was largely a pragmatic undertaking, with tried and true but self-limiting knowledge passed down from one generation of papermakers to the next. The science of papermaking was still in its infancy, with little real understanding of the chemistry of cellulose or the complexities of various paper properties. Papermaking was still a "feel-it, smell-it, taste-it" process.

But paper industry executives across the country began to recognize the need for concentrated research efforts within their own mills as well as within the industry as a whole. Several steps were taken in the early part of the 20th century, including the establishment of the USDA Forest Service's Forest Products Lab in Madison in 1910 and the development of the Technical Association of the Pulp and Paper Industry (TAPPI) and the American Pulp and Paper Mill Superintendents Management Association (PIMA) in 1915 and 1919, respectively. These organizations were formed to facilitate the exchange of scientific information and discuss possible solutions to the industry's technical challenges. Additionally, several notable research institutions, Syracuse University chief among them, began offering undergraduate courses in pulp and paper science. These efforts, though not well coordinated, were a vital step in the long path leading to the development of a graduate school fully devoted to the needs of the United States paper industry.

As paper industry executives pondered the challenge of advancing the industry in a technologically progressive age, a group of Wisconsin pulp and paper manufacturers came together to find a solution. The industry needed employees specifically trained in the scientific processes of papermaking. But who would train them? Individual mills lacked the resources needed to develop organized research programs and were unable to conduct a nation-wide search for men with adequate technical backgrounds to develop them.

In 1919, Dr. Samuel Plantz, President of Lawrence College, began to examine the possibility of organizing at Lawrence a pulp and paper curriculum similar to that at Syracuse University in New York. In April of that year, Plantz wrote a letter to Monroe A. Wertheimer, president of Thilmany Pulp and Paper Company and a trustee of Lawrence College, stating that "a school of this kind in our vicinity… would be of very great advantage to the men who are in the papermill industry." He did not envision a private graduate school of the magnitude that the Institute of Paper Chemistry would eventually become; rather, Plantz visualized undergraduate classes that would provide technical training to local mill personnel. Espousing the idea of adding a practical dimension to Lawrence's liberal arts curriculum, he wrote: "I should like very much to see the college relate itself more practically to the industries of our community."

Dr. Plantz's 1919 letter to Wertheimer outlined his plan in great detail. Lawrence would offer evening classes so that men who worked in paper mills during the day would be able to attend. He determined that a new building "with sufficient paper

I dedicate this institution to the highest and most efficient use of language and to the economical manufacture of the product which carries the knowledge of the ages to all mankind.

— **Dr. John H. Finley**, Associate Editor, *New York Times*, September 23, 1931 mill machinery" would be needed and that skilled professors must be hired to give instruction. He suggested hiring men from local paper mills to teach special evening classes while supplementing this instruction with a partial liberal arts curriculum to produce a well-rounded individual, or what would later be called a "scientific generalist."

Dr. Plantz knew that a business venture such as the one he was formulating would

be much better received were it to be presented by, as he told Wertheimer, "the initiative of some paper men like yourself." He further observed, "I feel quite sure if I were to go to the paper mill men of the valley, they would look at it as a scheme of the college to get a larger work developed and would perhaps doubt the efficiency of the enterprise."

The idea of a school devoted to the education and training of young men specifically for work in the paper industry began to spread among the ranks of Wisconsin's premier paper manufacturers. In a letter to Plantz dated July 6, 1920, L.M. Alexander, president of the Nekoosa-Edwards Paper Company, called the idea "very commendable because it is not only furnishing what we call a dependable job; but also at very good remuneration… and as Wisconsin is the third largest producing state in the United States and probably will be for some time to come – it is something that is worthwhile looking into."

It would be another nine years before the institution Plantz had in mind would become a reality, with delays centering on the source of funds for the school. Plantz had estimated that between $200,000 and $250,000 would be needed to construct a building, buy equipment and hire professors. Initially, none among Wisconsin's paper industry executives were willing to commit such an enormous amount of money.

Plantz remained an outspoken advocate of an institution dedicated to the training of young men for careers in the pulp and paper industry until his death in 1924. His immediate successor, Wilson Samuel Naylor, supported Plantz's plans, but his term as Lawrence's interim president proved too brief to act upon them. Then, in 1925, a professor of history at Connecticut's Wesleyan University was chosen by Lawrence's trustees to serve as the college's eighth president. Dr. Henry M.Wriston possessed, according to former Institute President John Strange in his 1979 publication *Fifty Years of Aspiration: An Abridged History of the Institute of Paper Chemistry*, "an uncompromising commitment to excellence" and though his tenure in Appleton was brief, his role in the founding the Institute of Paper Chemistry was significant.

Wriston did not act alone, however. His search among the ranks of Wisconsin's pulp and paper industry executives for a liaison, a man to link the technical with the educational and garner the support of his fellow "paper men," led Wriston to Ernst

Mahler, vice president and general manager of the Kimberly-Clark Corporation in nearby Neenah. Mahler was a Darmstadt man, distinguishing himself early on with his research in cellulose chemistry. His work caught the attention of Kimberly-Clark's top executives, who persuaded the Austrian to come to the United States and develop what would be one of the first research laboratories in the United States paper industry. Mahler's educational background – for it was intended that the new school model itself on the technical education philosophies of Darmstadt – combined with his experience and distinction in the industry, convinced Wriston that he would be the perfect candidate.

Wriston's choice of Ernst Mahler as the voice of the industry for the new school was a fine one. Mahler was both well-known and well-liked among paper industry executives of Wisconsin, and his voice was a strong one. By the early summer of 1929,Mahler delivered to Wriston a slate of nineteen pulp and paper companies – ninety percent of the paper manufacturers of Wisconsin – willing to support the new venture.

The Institute of Paper Chemistry was born.

Plans for the new school began in earnest.

Henry M. Wriston

In discussing the condition of liberal arts education in the United States, McGeorge Bundy, who, in addition to his role as an architect in foreign policy under the Kennedy and Johnson administrations, served as the Dean of the Faculty of Arts and Sciences at Harvard University, declared that, "we all go to school under Henry Wriston." His remark resonates with those fortunate enough to reap the benefits of Wriston's endless list of contributions to liberal arts education; indeed, Wriston certainly earned his title of the "American Educator."

Henry Merritt Wriston was born in Laramie, Wyoming, on July 4, 1889. He received his Bachelor of Arts degree in medieval history from Wesleyan University in 1911, followed by a Master's degree, also from Wesleyan, in 1912. Wriston pursued graduate studies in history and international affairs at Harvard University as an Austin Teaching Fellow, and in 1914, returned to Wesleyan as an instructor in history. In 1922, after Wriston had completed his dissertation and was awarded his PhD from Harvard, he was made a full professor at Wesleyan.

Wriston's leadership skills were evident even before his rise to several high-prole positions. As an undergraduate, he edited Wesleyan's newspaper, was a top debater, and won the senior oratorical contest. During World War I, Wriston served as the assistant manager of the Connecticut State Council of Defense. His success in this position led to his 1919 appointment as Executive Secretary of the Wesleyan Endowment Fund, which, in turn, greatly influenced Lawrence's Board of Trustees to select Wriston as the college's eighth president.

During his twelve-year tenure at Lawrence, Wriston earned national recognition for his numerous innovations in collegiate education, notable among them a program of educational and vocational guidance with experiments in adult education, art appreciation, and extensive reading programs. Upon Wriston's resignation from Lawrence in 1937, Appleton Mayor John Goodland acknowledged that the greatest strides in the history of Lawrence were made under Wriston's administration.

Most significant was his role in the founding of the Institute of Paper Chemistry. During Wriston's farewell banquet in late January of 1937, Ernst Mahler lavishly praised Wriston's vision, determination, and devotion in the Institute's development. "Indeed," he said, "the Institute has done more to put Appleton on the map than anything that has happened. The educational world had watched its growth and industries are fast realizing similar needs. Dr. Wriston, as director, has assembled most able educators, scientists, and administrators and leaves the personnel a solid foundation on which to build in the future... I believe Dr. Wriston will not stop as president of Brown University but is entitled by his abilities to the highest place the educational world has to offer."

Wriston left Lawrence in 1937 to become the eleventh president of Rhode Island's Brown University. The farewell he received was a testament to the admiration bestowed upon him by both Lawrence University and the town of Appleton – more than three hundred students and townspeople gathered at the Chicago and Northwestern train depot on the day of Wriston's departure to wave goodbye as the eastbound train pulled away from the station.

Scholar, writer, educator, administrator, political advisor – Wriston was truly a Renaissance Man. "He was a complex personage," noted James Adams, who served as Brown's vice president during Wriston's tenure as president, "but he had a brilliant mind, a steady nerve, a provocative manner, and an unwavering determination."

Ernst Mahler

Many men have the gift of technical skill, but inventive and imaginative insight are seldom added to it. In you these powers are brilliantly combined, and you have used them devotedly for the cause of international industry. But in these busy years you have also shown a connoisseur's passion for art and a scholar's passion for the cause of education.

— Lawrence College President Douglas M. Knight, in conferring upon Ernst Mahler an honorary Doctor of Law degree at the Institute's Silver Anniversary in 1954

These were the attributes that convinced Dr. Henry Wriston that Mahler was the right man to help bridge the gap between industry and education. Mahler did not disappoint.

Ernst Mahler was born in Vienna, Austria, in 1887 and raised in a world of paper chemists; by the age of twenty, he had left Vienna for Darmstadt, Germany, to attend the world-famous Technische Hochschule de Darmstadt. There he studied industrial chemistry, specializing in cellulose chemistry and the scientific study of papermaking. Shortly after his graduation in 1912, Mahler immigrated to the United States to begin work at the Kimberly-Clark Corporation in Neenah, Wisconsin.

Mahler quickly distinguished himself within the Kimberly-Clark Corporation. The research and development department that he established was among the first – and the finest – in the United States paper industry. Mahler led a team that was responsible for the development of many products still in use today. Perhaps most significant among Mahler's many contributions to the American pulp and paper industry was the development of creped wadding, a wood-based material that was used as a substitute for cotton bandages during the first World War. After the war, Kimberly-Clark turned Mahler's product into a multi-billion dollar industry, creating many new products that would eventually be trademarked under now-familiar brand names such as Kleenex and Kotex.

By the time that Wriston approached him regarding the vision of a school dedicated to the needs of the United States pulp and paper industry, Mahler had advanced to the position of vice president at Kimberly-Clark. The proposition was of immediate interest to Mahler, who for several years had been lamenting the trend toward vocationalism in colleges across the country. "Unfortunately the results of this trend were unsatisfactory," Mahler noted in a speech several years later. "By substituting 'commercial' or vocational courses for fundamental instruction, educators lost sight of the fact that true education is a study of values and not an imitative process."

Mahler's role in the founding of the Institute was immeasurable. He spent countless hours in boardrooms across the state, persuading Wisconsin's paper executives to provide the financial backing for the new institution. When representatives from the Institute's member companies met for the first time to discuss the organization of the school, Mahler's contributions were acknowledged by his fellow industry executives as they elected him president of the Board of Trustees. He maintained this position until the restructuring of the Board in 1950; afterward, he served as the Chairman of the Board until 1957. He was then named Honorary Chairman, a title he held until his death in 1967. During Mahler's tenure as the Board's leader, the Institute experienced enormous growth – new buildings and facilities valued at more than $4 million, an annual operating budget increasing from $40,000 per year in 1929 to $1.8 million in 1957, and a staff increasing from four members in 1929 to more than two hundred and fifty.

Mahler's contributions to his industry and his country were many and varied. He played a large role in the founding of the Technical Association of the Pulp and Paper Industry (TAPPI) in 1915 and was awarded the TAPPI Gold Medal for contributions to the industry in 1933. During World War II, Mahler served as a member of the Purchase Policy Division of the Army Service Forces, where he aided in formulating policies, reviewing legislation, and implementing the utilization of small war plants. In 1945, he spent five months in Europe as a consultant in charge of paper production and organization rehabilitation of the paper industries of Belgium, Holland, Germany, and France. For his services as a member of the United States Reparations Commission in Russia, Berlin, and Potsdam, Mahler was awarded a certificate of appreciation from the War Department and the Presidential Medal of Freedom. Among the many honors bestowed upon Mahler during his long and distinguished career were an honorary Doctor of Science degree from Brown University, an honorary Doctor of Law degree from Lawrence College, and an honorary Doctor of Engineering degree from his Alma mater, Darmstadt. These honors, according to a speech delivered by Dr. Henry Wriston in 1937, were quite betting one who was "a leader in transforming an ancient art through modern scientific processes."

In his 1982 treatise on the American free market system, *The Economy of Mind*, economist Warren Brookes wrote of Ernst Mahler:

Mahler was an entrepreneurial genius whose innovative ideas and leadership, over a period of about twenty years, transformed a once-small, insular newsprint and tissue manufacturer into one of the largest paper corporations in the world, which gives prosperous employment to more than 100,000 and produces products – which Mahler helped to innovate – that are now used by more than two billion people… his personal fortune was insignificant when compared with the permanent prosperity he generated, not only for his own company but for the hundreds of thousands who work for industries which his genius ultimately spawned and which long outlived him… the revolutionary sanitary products that have liberated two generations of women, or the printing paper that completely transformed international publishing and communications for fifty years.

The Institute's First Home: Appleton, Wisconsin

The Fox River Valley, in which Appleton nestles, has a rich and diverse history. First settled by the Winnebago, Menominee, Mascoutine, Patawatami, and Fox Indians, the area also became home to Jesuit missionaries seeking to convert the "savages" and capitalize on the Valley's rich fur trade. In 1836, the valley's four million acres was ceded to the United States by the Treaty of the Cedars.

In 1847, Amos A. Lawrence, a wealthy Boston merchant engaged in the manufacture and sale of textiles, found himself in possession of more than five thousand acres of Wisconsin wilderness that had served as collateral on a defaulted loan. Lawrence, who was well known for his philanthropy, commissioned three Methodist ministers to establish a school on his land. Lawrence pledged a $10,000 endowment for the school on the condition that the Methodist Church do the same.

The school's charter, granted in that same year, was radical for its time, promising religious freedom to its students, faculty and staff, and opening admission to men and women. It thus became only the second coeducational college in the country.

As Lawrence's representatives searched for a site for the school, it was quickly determined that his holdings were inappropriate, as they were dangerously close to the Indian and French settlements in the area. The committee turned its sights to the Fox River Valley and ultimately selected a spot at Grand Chute, the home of the White Heron Trading Post. Classes began at Lawrence Institute, the "university in the wilderness," in November of 1849. The town of Appleton, named after Samuel Appleton, Lawrence's father-in-law who gave $10,000 for the school's first library, developed around the small education institution, growing steadily from the crude huts of the early settlers to the beautifully designed homes and buildings of the present.

The choice of Appleton as the home for the Institute of Paper Chemistry some eighty years after classes began at Lawrence was decidedly strategic. Many early Institute publications and speeches boasted that practically every type of pulp and paper manufacture in the world could be found within a seven-mile radius of the city, giving Appleton a unique status in the industry. The town's proximity to these mills provided the newly formed Institute

with access to much-needed paper machinery and to highly trained men with practical papermaking experience, several of whom were engaged as lecturers and guest speakers.

Appleton, according to Ernst Mahler, vice president of Kimberly-Clark Corporation and a founding trustee of the Institute, was certainly "paper mill conscious." The picturesque town located on the banks of the Fox River proved a comfortable home for the Institute, and the two enjoyed a mutually beneficial relationship for more than sixty years.

FOUNDING TRUSTEES

LEWIS MILLER ALEXANDER

Nekoosa-Edwards Paper Company

DANIEL K. BROWN

Neenah Paper Company

DAVID CLARK EVEREST

Marathon Paper Company

ERNST MAHLER

Kimberly-Clark Corporation

HUGH STRANGE

John Strange Paper Company

MONROE A. WERTHEIMER

Thilmany Pulp and Paper Company

It was determined that the Institute would operate most effectively as a non-profit corporation and an independent entity, although it would maintain an affiliation with Lawrence College. The school would be governed by a Board of Trustees comprised of representatives from both Lawrence College and the Wisconsin paper industry in order to balance the academic and industrial missions of the Institute.

Financially, the Institute of Paper Chemistry would be supported by a membership structure. Companies producing pulp, paper, or paperboard would be eligible for membership and their dues – set at a yearly rate of $30 per ton of their daily paper production – would contribute to the school's operating budget. The original nineteen member companies brought in by Mahler pledged a total of $300,000 which, when augmented by funds from individuals and corporations allied with the paper industry, would guarantee the Institute's existence at a minimal operating budget for a period of five years.

The contributions of the Institute's member companies went above and beyond the financial. It was agreed that local mills would periodically send employees to the Institute to serve as guest lecturers. They also agreed to allow students access to their mills and provide them with hands-on experience not available in a classroom setting. Thus the pulp and paper mills surrounding Appleton became, in effect, laboratories for the Institute and its students.

Wriston, Mahler, and a handful of local paper executives dedicated to the school's development determined that the mission statement of the Institute of Paper Chemistry should focus on three specific functions. First and foremost, the school would provide postgraduate training to men who would then use that training to create and direct research laboratories in the nation's pulp and paper mills. The decision to develop the Institute as a graduate school was the subject of considerable debate throughout the summer of 1929. Other reputable institutions, such as Syracuse and North Carolina State, had well-established pulp and paper curriculums and the Institute's founders questioned the wisdom of simply duplicating those programs. It was ultimately decided that, though those undergraduate institutions were doing a fine job of instructing students in pulp and paper technology, the need for postgraduate training for men who had already completed their preliminary chemical or engineering training at the undergraduate level was immediate. Men with this training would be prepared to undertake the rigorous academic program the Institute would offer and, upon graduation, would ultimately possess the

qualifications to not only work in the United States' paper industry but to eventually guide its future as industry leaders.

The second purpose of the Institute would be to conduct research beneficial to the paper industry. As a result, the industry would greatly benefit from quality research and avoid duplication of staff and equipment. Most importantly, students would work alongside Institute staff, thereby receiving hands-on training in solving industry-related problems.

The founders determined that the Institute's third function would be to establish and maintain a complete library of pulp and paper science and technology. The library would support both the academic and research functions of the Institute by thoroughly covering the field and by providing abstract and translation services, benefiting not only Institute faculty, students and staff, but the entire industry as well.

As the fall of 1929 approached, the Institute's founders had one more item on their organizational agenda: developing an educational philosophy upon which to base the school's curriculum. Since the early 1900s, graduate education – especially in the fields of science and engineering – had been treading a path of specialization. Professors were focusing much more on *how* something should be done rather than *why* it should be done and, to quote John Strange, "higher education was in danger of becoming too vocational." The Institute's founders saw in this new school a chance to veer off that path and return to a more old-fashioned teaching philosophy still en vogue at some of the more prestigious academies in Europe, such as Technische Hochschule de Darmstadt (Darmstadt Polytechnic). "Europe has long recognized the necessity of such schools," said Mahler in a speech delivered at the American Pulp and Paper Association Annual Meeting, "and they are existing today in every industrial country over there."

The plan was simple. The Institute of Paper Chemistry would apply a liberal arts philosophy to the teaching of science and technology. Students would receive a broad education in the disciplines relevant to the manufacture of pulp and paper. Faculty would encourage imagination and creative thinking, grounded in a thorough knowledge of fundamental research procedures. This educational philosophy would, according to Mahler, "turn out a student… who can go out into the industry and create the confidence of his employer to the point where he can be trusted with the technical problems that arise."

At last all was in place. On October 12, 1929, Articles of Incorporation were filed with the Wisconsin Department of the State, seeking approval to organize a graduate school in affiliation with Lawrence College in Appleton. Within two weeks, the Articles had been granted and Wriston had called for a meeting to formally organize the Institute.

On Monday, October 28, the Lawrence library was filled with paper industry executives from throughout the state of Wisconsin, all eager to take part in this revolutionary venture. Representatives from each member company cast their votes and selected L.M. Alexander of the Nekoosa-Edwards Paper Company, Dan K. Brown of the Neenah Paper Company, D.C. Everest of the Marathon Paper Mills Company, Ernst Mahler of the Kimberly Clark Corporation, Hugh Strange of the John Strange Paper Company, and Monroe A. Wertheimer of the Thilmany Pulp and Paper Company as the founding trustees. Dr. Henry Wriston was named Director of the Institute.

After a quick lunch break, those assembled elected Mahler President of the Board of Trustees. Everest was chosen as the board's Vice President and Wriston became Secretary. Westbrook Steele, a Lawrence College employee considered an expert in educational structuring and financing, was appointed Executive Secretary and Ralph J. Watts, Lawrence's business manager, was named Treasurer.

The meeting adjourned in the early evening of what would later be known as Black Monday, the day the stock market suffered a record one-day loss of close to thirteen percent. Although the newly elected officers and their fellow paper industry executives were warily watching the events in New York unfold, they found solace in the progress made that day. They had secured several rooms in Lawrence's new Alexander Gymnasium to conduct lectures and labs once classes began the following February. They had set basic entrance requirements for prospective Institute students – a bachelor of science degree with a major in chemistry and a satisfactory personal interview. They had set the tuition rate at $200 per year, which would include all laboratory fees. Finally, they had given approval to begin the search for the Institute's first two employees – a librarian and a technical director.

The next day the founders and trustees – and the entire country – awoke to worsening news from New York. To most historians, "Black Tuesday" – October 29, 1929 – was a watershed in American history, marking the point where the Roaring Twenties ended and the Great Depression began. Although the market fell slightly less than the previous day (about twelve percent), a record 16.4 million shares were traded, causing the ticker tape to fall almost two and a half hours behind. By the end of November, investors had lost more than $100 billion in assets and the value of the stock market had dropped by forty percent. It would continue dropping for another two years, finally bottoming out in July of 1932 with the Dow at 41.22, down from 381.17 in the months before the Crash. It was not until 1954 that the stock market fully recovered.

"How easy and forgiving it would have been," John Strange wrote in his *Fifty Years of Aspiration*, "for the original sponsors to take the position that a storm was gathering and that it would be circumspect to delay their commitments until a more propitious time."

Yet they persevered – and they succeeded.

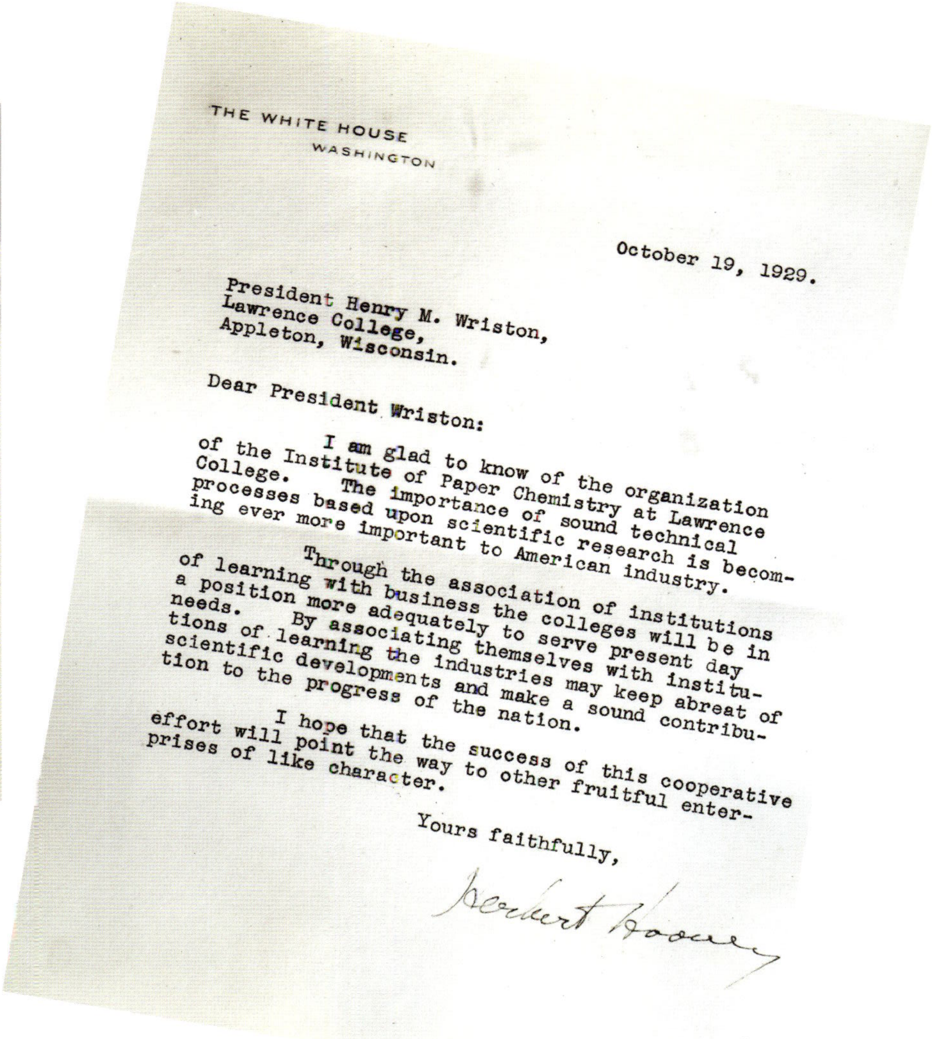

THE WHITE HOUSE
WASHINGTON

October 19, 1929.

President Henry M. Wriston,
Lawrence College,
Appleton, Wisconsin.

Dear President Wriston:

I am glad to know of the organization of the Institute of Paper Chemistry at Lawrence College. The importance of sound technical processes based upon scientific research is becoming ever more important to American industry.

Through the association of institutions of learning with business the colleges will be in a position more adequately to serve present day needs. By associating themselves with institutions of learning the industries may keep abreast of scientific developments and make a sound contribution to the progress of the nation.

I hope that the success of this cooperative effort will point the way to other fruitful enterprises of like character.

Yours faithfully,

Herbert Hoover

CHAPTER 1

1930-1939

The world awakened to a solemn dawn in 1930. Even as politicians and the leaders of great American industries tried to sooth the dazed country with optimistic predictions and promises of "a brighter tomorrow," banks, factories, and thousands of other businesses closed their doors, leaving millions of Americans jobless and, in many cases, homeless and hungry. The American Dream had become, for many, a seemingly endless nightmare.

But in a small midwestern town on the banks of the Fox River, the dreams and visions of a group determined to shift the direction of higher education in the United States were quickly being realized...

"The Institute of 1930," remarked Harry Lewis in his August 1939 newsletter, " was largely Otto [Kress] at one end of the log and three students at the other."

WHAT MADE THIS COMMENT so humorous to anyone familiar with the Institute's humble beginnings was the element of truth it contained. Yes, there were a few – very few – other faces besides that of Dr. Otto Kress floating around in those early days. There were more than three students, but they were "special students" who worked in local paper mills and attended classes only part-time. And the few rooms in Lawrence's Alexander Gymnasium, modest though they were, were still more than simply a log. However, when compared with what the Institute would become, the mental picture Lewis's comment evoked did resonate, especially with those who had been around long enough to experience that growth first-hand.

Dr. Otto Kress

Wriston,Mahler, and the other trustees elected at the October 28 board meeting wasted no time following up on the plans made that day.Memos flew back and forth from Appleton, Neenah, and surrounding areas as the Institute's founders debated course offerings, scholarship monies, administration, and a host of other issues associated with the development and management of a graduate school.

Two positions needed to be filled immediately. The first was technical director. The position called for, as an early Institute academic brochure explained, "a man not only broadly trained in the various fields of pulp and papermaking but for one who also possessed wide contacts among the technical men associated with the paper and allied industries." No man fit the job description more perfectly than Dr. Otto Kress, who had been educated at New York's Columbia University and possessed more than twenty years of practical experience in the field of pulp and paper. It was, as Mahler once said, Kress's immediate and enthusiastic willingness to be the technical director, the only full-time teacher and the general factotum - including millwright and janitor - that formed one of the cornerstones of the Institute and made Kress a legend.

I had the privilege of visiting the Institute to recruit young engineers into the paper industry. I would go out to the Institute twice a year and talk to the boys out there. I was always impressed by the way new buildings came up and what started out as two buildings became three, four, five buildings. I was also impressed that they were still doing a good job with their training.

— Daniel Adams, entering class 1939

It was a happy time for me.

— Al Schenck,
entering class 1938

I learned a lot. In essence, they made me a scientific generalist.

— Edwin Loutzenheiser,
entering class 1939

Institute faculty meeting, c. 1931

The second position that required immediate attention was librarian. One of the three primary functions of the Institute of Paper Chemistry as discussed during the October 28 meeting was the establishment and maintenance of a complete library dedicated to the many aspects of pulp and paper science and technology to benefit the Institute students, faculty, and staff as well as the United States paper industry as a whole. On April 21, 1930, it was announced by Westbrook Steele, who was quickly becoming the Institute's chief promoter, that Hjordis Roseth had been chosen to fill the position.Miss Roseth, originally from Oslo, Norway, had attended New York State Library School in Albany, and at the time of her IPC appointment was working at the Riordan Research Institute of the Canadian International Paper Company. Although Miss Roseth only served as librarian for one year, it was she who first organized and arranged what would become the most comprehensive collection of pulp and paper reference materials in the country.

It is worth noting again that creating a "normal" graduate program is almost always difficult at best. To create a program the likes of which had never before been seen in the United States made the task all the more complex. A case in point was the hiring of faculty and staff as February 3 – the date classes were scheduled to commence – drew near. According to Mahler, there were three possible courses of action that could be taken with regard to the hiring of faculty members: "The first is to secure a group of academically trained men, who are familiar with the processes of teaching and research in pure science. A second would be to take chiefly men of middle age who have long experience in the problems of productions but who have slight experience in the problems of teaching. A third would be to take a group of brilliant young men, just out of graduate school, and train them for our purposes." The beauty of developing such a unique institution was at once obvious. No one particular path was essential; the Institute's founders could, as Mahler suggested, "choose all three." And so they did. With the Institute's very first matriculation day so close at hand, little time was left to conduct a search for the type of men Mahler,Wriston, and others were looking for in an IPC professor and researcher. Since it was inconceivable that Dr. Kress could teach all the classes on the proposed syllabus, it was suggested that the Institute "borrow" several professors from Lawrence College to ease Kress's teaching load. Lawrence professors Dr. Gottlob Cast, Dr. Stephen Darling,Mr.Walter Rogers, and Dr. Lewis Youtz answered the call. Although the positions were originally temporary, each of these four found a niche at the Institute of Paper Chemistry and eventually rounded out their prestigious careers as professors and researchers at the new institution.

Cast was recruited to teach a class in Technical German, deemed necessary due to the number of technical journals written in German.William Haselton, entering class 1949, recalls Cast as "strictly from the old school" but added that as long as you did your work, you would get along fine with him. George Ingle, entering class 1938, remembers with a laugh the difficulties

Technical German Class with Dr. Cast, c. 1931

"the guys from the southern schools who had never had a word of German in their lifetimes" had with Cast's course. "I tried to help them," he recalls. "But they were in a tough situation."

Darling, who had received his doctorate from Harvard University in 1928 and thus was not much older than many of his students, was brought in to teach several areas of organic chemistry, including the chemistries of cellulose, carbohydrates, and lignin. Ingle, who describes Darling simply as "a likeable guy" remembers the fun times he and his fellow classmates had in the young professor's lab extracting rare sugars from bales of various weeds and plants. "I think he sold them [the extracted sugars] and made some money for the Institute," Ingle said.

Rogers, who had spent much of his career in the paper industry, taught courses in wood chemistry and often accompanied the new entering classes on their bush trips, a facet of Institute life introduced in September of 1930. Although he and Youtz, who was hired to teach technical analysis, did not spend quite as much time at Institute as did Cast and Darling, this Lawrence quartet removed a great deal of the teaching burden from Kress, who taught a total of four classes that first semester, and Miss Roseth, who taught a short course in technical abstracts and bibliography.

I remember it as a pleasant period in my life. I was busy and glad to be there, working to get my degree.

— Roy Davis, entering class 1930

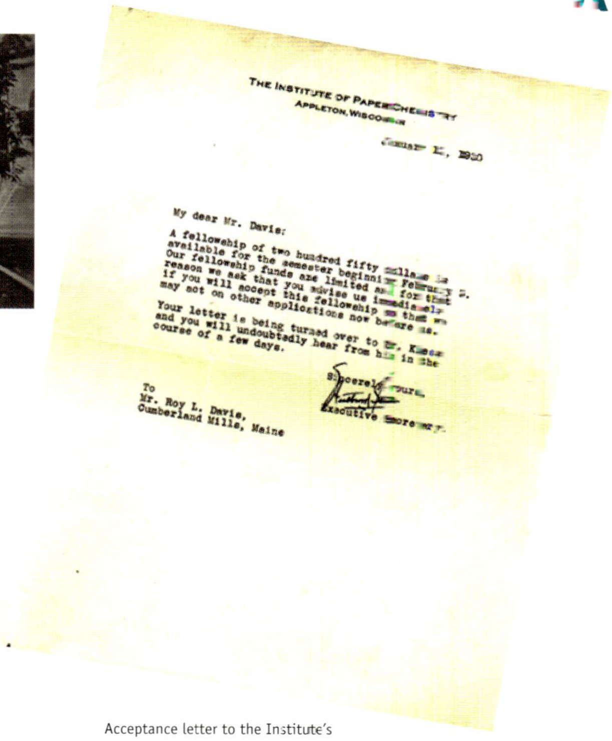

THE INSTITUTE OF PAPER CHEMISTRY
APPLETON, WISCONSIN

January [illegible], 1930

My dear Mr. Davis:

A fellowship of two hundred fifty dollars is available for the semester beginning February [illegible]. Our fellowship funds are limited and for that reason we ask that you advise us immediately if you will accept this fellowship so that we may act on other applications now before us.

Your letter is being turned over to Dr. Kress and you will undoubtedly hear from him in the course of a few days.

Sincerely yours,

Executive Secretary.

To
Mr. Roy L. Davis,
Cumberland Mills, Maine

Acceptance letter to the Institute's first entering class, 1930

At last the long-awaited day arrived. On Monday, February 3, 1930, coinciding with the commencement of Lawrence's second semester, Harold Bialkowsky, Roy Davis, and Ed Voightman, along with several other part-time men from local mills, joined Kress in a large room positioned directly over the lobby of the Alexander Gymnasium, where, according to Lewis, "any convenient tabletop, first cleaned of acid or alkali, formed the chair of the teacher, and other similar perches the resting places of the students." Roy Davis recalls not being fazed by the lack of available lab space or the makeshift laboratory equipment, much of which was developed by Kress himself."We just accepted it," he stated good-naturedly. "That was the situation.We were getting the courses we needed."

That first semester was, by all estimations of those involved, truly an experiment in education. The students were guinea pigs, cheerfully adjusting to changes in the curriculum, the crude equipment, and the living conditions of Brokaw Hall, the Lawrence College freshman dormitory that housed the first three Institute students at a rate of $350 per year.

Dr. Harry Lewis

Those students, according to Davis's recollections of that first semester, never minded Kress having to leave the classroom in order to answer a ringing phone or to clean up a mess in one of the labs. In fact, the students themselves often pitched in as well, doing their share of clean-ups, secretarial work, and library assistance. Perhaps it was the makeshift environment, or the double and sometimes triple-duties performed by Kress and others, or just simply the knowledge that something grand was happening, but, as Lewis later noted, "the best teaching the Institute ever had was done that first semester beginning February 1930."

Construction of the original General Activities Building, 1931

The first semester of classes at the Institute of Paper Chemistry, though hailed as an enormous success by all standards, was also a learning experience, carried out, according to Lewis, "more or less as a pilot plant study is utilized in the mill prior to a full-scale installation." As the special students returned to their respective mills and the three full-time students left for their summer jobs,Wriston,Mahler, and Kress began the process of evaluating progress made during the previous four months. There was, it was agreed, much to be done before the new class of students – a full class this time – came to Appleton in September.

The first order of business was to hire another faculty member. A graduate school needed a dean, it was decided. But for the Institute's purposes someone with only administrative skills would not suffice. The Institute needed a man who could aid Kress in the administration of the program, who was familiar with the paper industry, and who had some teaching experience under his belt. Kress found everything the newly formed graduate program needed – and a great deal more – in Dr.Harry Fletcher Lewis.

Lewis, who had received his PhD from the University of Illinois in 1916, joined the Institute staff as a professor of organic chemistry in June of 1930.Within a year, the Board of Trustees voted unanimously to appoint Lewis the dean of students; in Lewis's own words, "the budding organization needed someone who knew the educational language and rules and at the same time had some acquaintance with the descriptive phrases of the technical man." As dean of students, Lewis found himself in charge of admissions, academic organization, and supervision of the academic work undertaken at the Institute. It was, as it turned out, a job for which he was aptly suited. Like Kress, Lewis was soon recognized as a cornerstone of the Institute and, also like Kress, eventually achieved legendary status.

The summer of 1930 passed quickly and on September 15, 1930, seventeen students - graduates of fifteen colleges and universities across the country - met in Appleton to begin the first full year of classes at the Institute of Paper Chemistry.

Otto Kress

"In him, there is a combination of ripe technical experience, an interest in students and an undying interest in anything at all that relates to pulp and paper."

- Dr. Harry F. Lewis, Dean of Students November 2, 1934

"I envy those young men who have the privilege of sitting at your feet. Through them you are making an impression on this grand industry, giving it a shape that will last for many years and form one of the important chapters in the history of pulp and paper."

- Mr. Ernst Mahler President, IPC Board of Trustees TAPPI Medal Presentation February 21, 1940

Born in 1884, Otto Kress received his PhD from New York's Columbia University in 1909. By the time he came to the Institute in 1930, he had gained more than twenty years of practical experience in the paper industry with positions at the Forest Products Laboratory at Madison, Badische Pulp and Paper Company, Consolidated Water Power and Paper Company, and Thilmany Pulp and Paper Company.

Known to his students as "Herr Professor," Kress was an invaluable contributor to the Institute's formation and development during its adolescent years. His colleagues and students often wondered why Kress, who was once described by Henry Wriston as a "genius in the paper industry," would leave his good position within the industry to join forces with the new and untested Institute of Paper Chemistry. During his acceptance speech of the 1940 TAPPI Gold Medal, Kress answered them: "It is this work in which I am presently engaged that I have found to be so satisfying, as the opportunity is presented not only to help in conducting suitable research work but, more important, to pass on to a limited group of intelligent men and interested young men the information which had been accumulated through the past ages and thus fit them for a career in their chosen field of pulp and paper manufacture."

Kress retired from his position as the Institute's Technical Director in 1948. He remained active in Institute life until his death in 1970, notably as the donor of more than $70,000 to a scholarship fund aptly named the Otto and Florence T. Kress Scholarship Fund.

Only a man of the vision of Otto Kress could have left his good position and helped to found the Institute. Other men could have given perhaps as much to the students and to the industry in the world of knowledge but no one else could have imprinted his character on the Institute and the students during the years like Otto Kress. As the graduates assume more and more executive responsibility in our industry, they and we owe an everlasting debt to Otto Kress.

- Ernst Mahler, President, IPC Board of Trustees, June 5, 1948

Otto Kress, we live in a part of the world where pioneers and pioneer virtues are held in high esteem. Pioneer member of a pioneer enterprise, the Institute of Paper Chemistry, before that, one of the pioneer doctors of philosophy to specialize in the problems of the pulp and paper industry, you have recently and, from our point of view, regrettably, pioneered again in becoming the first emeriti of the Institute you have so ably and loyally served. Inventor, teacher, equally at home in theoretical and applied aspects of chemical knowledge – in laboratory, classrooms, and the mill – you have been a great credit both to the Institute and to the pulp and paper industry.

- Dr. Nathan M. Pusey, President, Lawrence College, June 24, 1948

As those seventeen men settled in to the rigorous regime of the typical Institute of Paper Chemistry student, Wriston, Mahler, and Kress turned their attentions to a new problem: space, or, rather, the lack thereof. The makeshift laboratories and offices in the rooms on the second floor of Alexander Gymnasium were quickly becoming crowded and the library, which had grown enormously the previous year, was quickly running out of storage space. The answer was simple: the Institute of Paper Chemistry needed a home of its own. Construction began in the spring of 1931.

June 8, 1931, was an epic day for the Institute. During Lawrence College's 81st Annual Commencement Ceremony, ten men - Edward Amstutz, Harold Bialkowsky, Karl Buff, Wilbur Cassiday, Richard Crain, Roy Davis, Emanuel Foreman, Harry Irwin, Edwin Laughlin, and Carl Schroeder – were granted the first Master of Science degrees awarded by Lawrence College for graduate work completed at the Institute of Paper Chemistry. Later that afternoon, faculty, staff, and students from both institutions gathered in a large field across the street from the Alexander Gymnasium to witness the laying of the cornerstone of what would become the Institute's permanent home.

Miss Edith Stroschneider

As the first full school year came to a close, the Institute experienced its first staff rollover as the librarian, Miss Roseth, announced her resignation and her intention to return to Europe. She was replaced by Miss Edith Stroschneider, who, throughout the course of her twenty-fiveyear tenure at the Institute, brought the library recognition as one of world's most complete collections of materials relating to the production of pulp and paper.

Born in Vienna, Austria, in 1889, Miss Stroschneider attended the Studley Horticultural Institute in Birmingham, England. She came to the Institute in 1931 from the Abitibi Power and Paper Company in Canada, where she served as a librarian. Her experiences at Abitibi gave her a wide range of knowledge and a great appreciation of the paper industry, which proved quite useful as she commenced her duties at the Institute. Described as "intensely interested in her work," Miss Stroschneider held a myriad of responsibilities at the Institute, ranging from the organization of the library's materials to assisting students with research to teaching the course in Technical Bibliography. Most prominent among her duties, though, was the editorship and monthly production of the Abstract Bulletin, a forty-page publication listing recent additions to the library's collection and a compilation of abstracts of all important articles from the library's subscription list that was sent to all member companies and pulp and paper mills that provided support to the Institute.

Miss Stroschneider possessed unique talents as well as certain idiosyncrasies that made her one of the more memorable characters of the Institute's rst quarter-decade. She spoke seven languages well and had a rudimentary knowledge of several more, which allowed her to decipher the many technical publications published in a wide range of languages that came across her desk in the J.A. Kimberly Memorial Library. Her dedication to her work was immeasurable; a favorite expression around the Institute throughout the 1930s and 1940s (akin to the modern "when pigs fly") was "...when Miss Stroschneider takes a vacation."

Miss Stroschneider officially retired from the Institute in 1953 and was graced with the title of "Emerita," though she continued to serve as the editor of the Bulletin until her death in 1956. Dr. Lewis honored her best in his February 1957 newsletter:

> "No one will have contributed more to the day-to-day program as well as to the permanent reputation of the Institute than Edith Stroschneider."

The librarian was very enthusiastic and she was very helpful in every way. Any information we wanted she knew where to find it.

— Joe Atchison, entering class 1938

The librarian, Miss Stroschneider, taught me a great deal – a real good old German gal. She was an excellent librarian. She taught me things about technical libraries that carried me through my days in Washington.

— George Ingle, entering class 1938

Edith Stroschneider - she was in charge of the library. She ran a tight ship in the library.

— Walt Bublitz, entering class 1941

I remember the librarian's little black poodle who spent the day in a basket under her desk.

— Ben Haug, entering class 1941

The day of the building's dedication – Wednesday, September 23, 1931, which coincided with the matriculation of the Institute's second entering class of students – was equally significant. The new General Activities Building, built at a cost of $250,000, was an architectural masterpiece. Designed by Orbison and Orbison Engineers, the building was two stories high with two deep wings attached.
One wing was designed to house the thirty-foot by sixty-foot digester room used for the cooking of wood and rags and the production of pulp from the half-stock on the first floor and a variety of special labs on the floor above, including a large lab for strength testing and color matching and another for microscopic study of fibers and finished paper. The other wing was devoted to work in colloid chemistry, containing a small research lab and a large student lab as well as a dark room for optical work and another for photographic work. The first floor of the building's front section contained the Institute's offices and classrooms as well as the rapidly expanding library; the second floor consisted of two large student research labs for work in the general field of pulp and paper technology, a large student organic lab and weighing room, and a small research lab for work in cellulose chemistry.

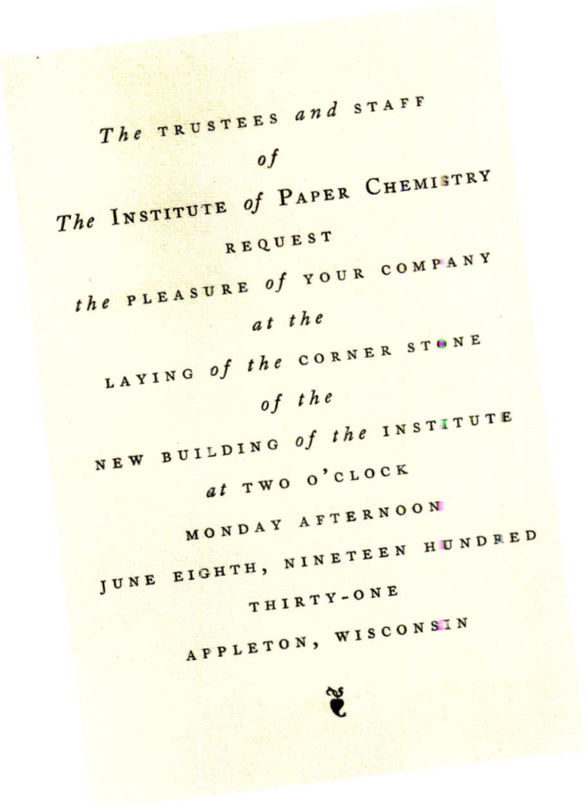
The TRUSTEES and STAFF
of
The INSTITUTE of PAPER CHEMISTRY
REQUEST
the PLEASURE of YOUR COMPANY
at the
LAYING of the CORNER STONE
of the
NEW BUILDING of the INSTITUTE
at TWO O'CLOCK
MONDAY AFTERNOON
JUNE EIGHTH, NINETEEN HUNDRED
THIRTY-ONE
APPLETON, WISCONSIN

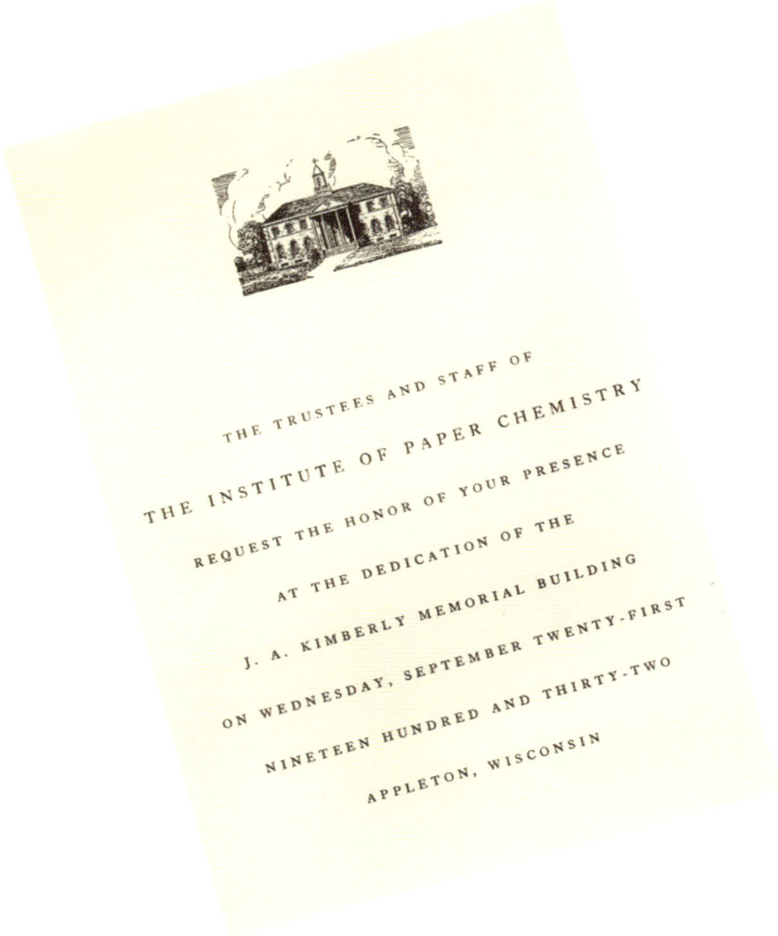
THE TRUSTEES AND STAFF OF
THE INSTITUTE OF PAPER CHEMISTRY
REQUEST THE HONOR OF YOUR PRESENCE
AT THE DEDICATION OF THE
J. A. KIMBERLY MEMORIAL BUILDING
ON WEDNESDAY, SEPTEMBER TWENTY-FIRST
NINETEEN HUNDRED AND THIRTY-TWO
APPLETON, WISCONSIN

The Institute's First PhD class, 1933

Contributing to the excitement of the day was a special announcement made by Mahler during the key presentation ceremony. J.C. Kimberly, president of Neenah Paper Company, had donated $100,000 for a memorial to honor his father, Wisconsin pulp and paper magnate J.A. Kimberly who, after serving as president of the Kimberly-Clark Corporation for 56 years, had passed away in 1928. The gift was intended, according to Mahler's speech, "to erect contiguous to the present building, a library and laboratory to house the splendid collection of books we possess and to make provision for its growth and to provide additional laboratory space already shown desirable." Construction of the J.A. Kimberly Memorial Library began immediately.

Likes its sister building, the Kimberly Library was an architectural magnum opus. Resembling the General Activities Building with an exterior entirely done in Lannon stone, the building, in a classic colonial architectural style, was designed with a high-pillared porch and a cupola with a clock. Antique pine paneling covered the walls of the first floor, which hosted a spacious reading room, a recreational reading room, the librarian's office, and a stock room. The colloid chemistry labs, previously in the General Activities Building, were relocated to the second floor of the new building, along with a series of conference rooms. The building's basement contained optical and photographic dark rooms, a large optics lab, a research reading room, and the librarian's workroom.

The Institute's First Masters class, 1931

By the time the Institute's third anniversary rolled around, thirty-five students were attending classes in the General Activities Building. Seventeen of these students were in their first year, ten in their second year, and eight in their third year. The Institute's annual budget had grown to more than $100,000 and six paper companies had joined the Institute's membership, bringing the total to twenty-five. A press release circulated during the anniversary celebration in February of 1933 told the member companies that:

> *The aims of the Institute have been followed with some degree of success. Mills and individuals responded generously, and, as a result, buildings and equipment have been forthcoming. The reputation of the Institute is constantly spreading and the interest in it widening geographically. The foundation seems stable – its future, hopeful. It is especially gratifying to have found such a response to its aims in the midst of depression that has done so much to deaden activity in general.*

On Monday, June 12, 1933, four of the "stones" comprising the stable foundation upon which the Institute was built made Institute – and Lawrence College – history by receiving the first Doctor of Philosophy degrees ever granted by the college. One by one, Harold Bialkowsky, Roy Davis, Edwin Laughlin, and Ed Voightman took the stage during the commencement ceremony to be conferred by Dr.Wriston, presented by Dr. Lewis, and invested with a hood by Dr. Kress, thus culminating three and a half years of intense training in pulp and paper technology. The Institute was on its way.

I'd say it's probably outdone Westbrook Steele's early vision of being a force in the industry.

—John McEwen, entering class, 1939

It was a good memory – an interesting period in my life.

—Joe Atchison, entering class 1938

The 1930s was a decade of colossal change; a dynamic situation heartily embraced by the Institute. Curricula were created, tested, revised, and tested again as the Institute's academic advisors searched for the best and most thorough way to fill a man's head with the knowledge he needed to become a major force in the United States pulp and paper industry. The original nine courses had expanded to more than thirty by the decade's end. A new division was created to keep up with the industry's emerging emphasis on physics.Work in fiber microscopy and photomicrography was emphasized, and the scope of work in pulp and paper testing was greatly enlarged. Entrance requirements were tightened as more and more students throughout the country became familiar with the Institute and desired admission. By 1939, more than two thousand applications for admission had been received.

Dr. Ben Rowland

Harry Fletcher Lewis

Harry Fletcher Lewis was born on May 27, 1891, in Denver, Colorado. He spent his childhood and teenage years in Manhattan, then moved to Connecticut to attend Wesleyan University. Upon receiving his Master of Science degree from Wesleyan in 1913, Lewis transferred to the University of Illinois, where he received his PhD in organic chemistry in 1916.

Lewis held a variety of positions before arriving in Appleton to become dean of the Institute. He was a member of the research sta of the U.S. Department of Agriculture Bureau of Chemistry; then in 1919, he joined the research department of the National Analine and Chemical Company in Bualo, New York, where he aided in the development of the country's post-World War I dye industry.

He taught organic chemistry at several colleges and universities, including Iowa's Grinnell and Cornell colleges, the University of Maine, and Ohio Wesleyan University. He was serving as head of the department of chemistry at Ohio Wesleyan when he left that school to join the sta of the Institute of Paper Chemistry in 1930.

Though his duties as dean kept him busy, Lewis remained a scientist at heart and still found joy in his research in a variety of elds, including redwood chemistry, lignin and its conversion to useful products, the permanence of paper, chemistry of bark, and the structure and property of vegetable tanning. Cellulose chemistry remained his great passion; he was once quoted as saying, "the only denition one can give is that cellulose is a happy idea towards which everyone works but at which nobody ever arrives."

Lewis's career at the Institute was lled with awards, honors, and recognitions. He received two honorary doctor of science degrees, one from Ohio Wesleyan and one from Grinnell College. In 1955, he was awarded the Mitscherlich Medal, an award presented by the Vereinzellcheming, an association of pulp and paper chemists and engineers in Germany. At the time of the award, Lewis held the distinction of being the only Mitscherlich Medal winner in the history of the award who was not in possession of a European education. In 1956, Lewis received the United States equivalent of the Mitscherlich Medal, the TAPPI Gold Medal, which was presented to him by one of his very rst students at the Institute, Dr. Roy Davis.

In late 1956, at age sixty-ve, Lewis announced his retirement from the Institute of Paper Chemistry. He remained active in Institute life for several more years, serving for a period as the Institute's vice president, then, upon receiving emeritus status, working with Dard Hunter in the Dard Hunter Paper Museum as a curator. An article in the June 1956 issue of the Insti-Tooter summed up Lewis's retirement period perfectly: "to speak of him as being 'active' is weak – he will continue to serve as crusader."

He was active and he knew what was needed for this Institute to become a recognized educational institution.

— Roy Davis, entering class 1930

He was kind of a weird character, I thought, but at least he enthused me.

— George Ingle, entering class 1938

He was a very strong personality around the school. I don't recall him teaching, but he certainly was very evident and very much a part of the program.

— George Mead, entering class 1950

Harry Lewis was a small guy. In our class of fteen people, I think every one of us was bigger than he was.

— John McEwen, entering class 1939

He was a bubbly, cheerful guy. Always upbeat.

— Al Schenck, entering class 1938

Harry Lewis was a wonderful guy. He really cared about the students, as well as a number of companies and the quality of education and everything.

— Arthur Dresheld, entering class 1951

A steady stream of the country's brightest scientific minds came to Appleton to educate those fortunate enough to gain admittance. In December of 1930 Dr. Ben Rowland, who had received his PhD from the University of Wisconsin in 1925, joined the staff as a research associate and professor of colloidal chemistry. According to Dr. Lewis in a speech made some years later, "Dr. Rowland came in and overwhelmed them [the students] with the language of colloidal chemistry." Dr. Bertie L. Browning joined the staff in 1933 as an assistant to Lewis on the Rag Content Project; he became a full staff member in the analytical department a few months later. Two Germans, Dr. Friedrich (Fritz) Brauns and Dr. Emil Heuser, came on board in 1935 and 1938, respectively, and others – Dr. Johannes Van den Akker, Dr. Clarence J. West, John Graff and Sid Wells, to name just a few – rounded out the academic team.

In October of 1936, all in Appleton were dismayed to hear that Dr. Wriston, who had been so instrumental in the founding of the Institute, was leaving to assume the

Dr. Henry M. Wriston

presidency of Rhode Island's Brown University. Luckily, the Lawrence College board of trustees named as his successor Dr. Thomas Nicholas Barrows, who had served as Lawrence's dean since 1934 and who was very familiar with the Institute and its programs. According to a speech given by Mahler at Barrows' inauguration, "Under the guidance of Dr. Barrows, the educational program of the Institute will continue to be one of real substance and the balance between industry, research, and education will be maintained."

The J.A. Kimberly Memorial Library also grew by leaps and bound during the Institute's first decade. Edith Stroschneider, who took over the position of librarian after Miss Roseth resigned at the end of the 1930-1931 school year, often

Dr. Thomas N. Barrows

boasted of the completeness of the library's collection, which, by the Institute's fifth anniversary, contained some sixtyfive hundred bound volumes and subscribed to more than one hundred technical and scientific journals, fifty percent of which were foreign language publications. The Abstract Bulletin, a forty-page bound volume listing recent additions to the library's collection and a compilation of abstracts of all important articles from the library's subscription list, was published monthly and sent to all member companies and pulp and paper mills that provided support to the Institute.

As the drums of war were sounding in far-off Europe, the Institute of Paper Chemistry wrapped up its first decade. As Lewis said in his June 1940 newsletter, "In spite of wars and depressions, we can truthfully say that these ten years have been eminently successful in every respect...the wisdom of the organization of the Institute, its aims, and its programs, has been confirmed many times over."

The Summer Job Program

In an early memo detailing the organization of the Institute of Paper Chemistry, Ernst Mahler developed a proviso requiring Institute students to forego their traditional summer holiday:

"During the summer months, the students are required to work in a pulp or paper mill, and in case they have not come to the Institute from other mills, positions will be secured for them at one of the local mills. In case, however, they have come from mills located elsewhere, it is to be expected that they will return to those mills during the summer months."

The Summer Job Program was born.

The program grew along with the Institute. In the earliest days, summer mill work was limited to upper classmen, but by the late 1930s, first-year students were participating as well, as faculty members stressed the importance of gaining as much practical experience as could be fitted into the four-year curriculum.

Here, Bob Miller, entering class 1936, describes his summer job experience:

The summer of my first year at the Institute, I worked for the Schmidt and Ault Paper Company in York, Pennsylvania. They made a variety of box-type stock on a cylinder machine, paten coated whites, etc.. They used a great deal of waste paper among which were the usual telephone books. When I wasn't running tests, I used to shoot the breeze with the beater operator. We would stand on a platform from which the beater was loaded. The favorite pastime was to feed the telephone books tossing them in. I gradually worked up to those from New York City, which were sometimes two inches thick. Try it sometime. There is a trick, of course, by flexing the book at the spine.

The second summer was at a similar mill in the Midway section between Minneapolis and St. Paul. That was a long, hot summer. I recall working behind the calendar stack of a paper machine feeding wax emulsion to a long tray that wet the paper, which was then ironed by the lower calendar rolls. The ironing action was achieved by the differential speed between two rolls. The friction set up a high-pitched screeching, which was deafening. This was before OSHA and ear plugs. My ears rang for a week after. Later, when I enlisted in the Navy, I barely made the hearing test.

The J.A. Kimberly Memorial Library

In September of 1852, as the story goes, a young John Alfred Kimberly from Neenah paid tuition of $6.67 to become one of the very first students to matriculate at Lawrence University. From Lawrence, J.A. Kimberly moved easily into a sixty-yearlong career as one of the most successful pulp and paper manufacturers in the United States. Co-founder of the Kimberly-Clark Corporation and operator of a number of mills in Neenah, Kimberly, and surrounding areas, J.A. Kimberly was, according to Ernst Mahler, "a pioneer in developing research into the technical problems of the industry."

On September 23, 1931, almost eighty years after Kimberly entered Lawrence University, the Institute of Paper Chemistry, a graduate school devoted exclusively to the industry to which Kimberly had dedicated his life, formally opened the doors of its first building. It was fitting that Mahler chose that particular occasion to announce a special gift, the largest received by the Institute thus far: $100,000 from James C. Kimberly, J.A. Kimberly's son and president of Neenah Paper Company, to construct a memorial to his father. The resulting building, hailed by Henry Wriston as "an illustration of the great moral impulse which has made American philanthropy the wonder of the world," would represent the first structure named in honor of a Lawrence alumnus.

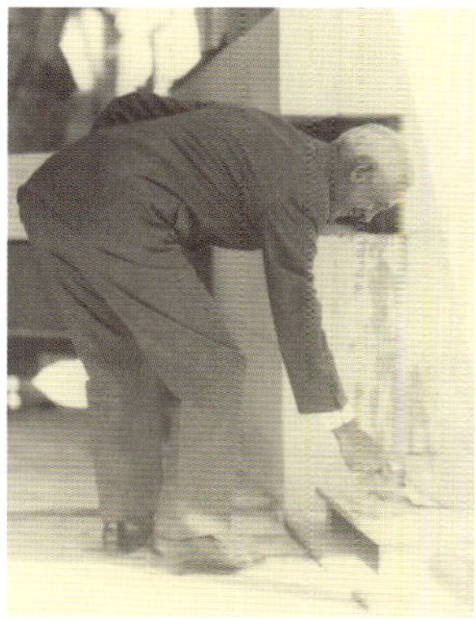

Construction of the new library, which was designed by Chicago architect Chester A. Walcott, began in the spring of 1932. On June 23, 1932, members of the Kimberly family joined the IPC and Lawrence communities, as well as more than two hundred industry executives, in a ceremony celebrating the laying of the building's cornerstone. Placed in the cornerstone box were the IPC dedication booklet, several copies of the IPC Bulletin, J.C. Kimberly's letter confirming the $100,000 gift, a program from Lawrence College's 1932 commencement exercises, and IPC course catalogues from 1930 and 1931. Kimberly himself laid the cornerstone, stating that, "it is the purpose of the Institute, through scientific methods and research, to raise the standard of the paper and pulp and allied industries to a higher level; the education of the student is a necessary means to this end. Without efficient manpower no industry can progress."

The J.A. Kimberly Memorial Library was dedicated on September 21, 1932, representing, as Wriston declared in his matriculation day speech, "the scientific progress of a great industry toward the development of which he [J.A. Kimberly] made great contribution by the application of his energy, his intelligence, and his appreciation of scientific work."

The Institute's First Semester
Course Catalog

COURSE 1:
INORGANIC TECHNICAL ANALYSIS

Professor Youtz

DESCRIPTION: Methods of analysis and interpretation of the data obtained by the experimentation of important raw and intermediate materials used in the pulp and paper industry.

HOURS: 2 hours lecture a week, 2 afternoons laboratory a week.

REQUIRED TEXTS:

- *The Manufacture of Pulp and Paper*
  Volumes III, IV, and V
- *Chemische Betriebskontrolle*
  by Schwalbe-Sieber

COURSE 2:
CHEMISTRY OF CELLULOSE AND ALLIED PRODUCTS

Dr. Darling

DESCRIPTION: The introductory course covers a discussion of the chemistry of carbohydrates, their structure, and reactions. This is followed by the chemistry of cellulose from the standpoint of its known chemical reactions and a similar study is included of lignin.

HOURS: 2 hours lecture a week, 1 afternoon laboratory a week.

REQUIRED TEXTS:

- *Textbook of Cellulose Chemistry*
  by Emil Heuser
- *The Chemistry of Wood*
  by Hawley and Wise
- Standard Methods of Testing
  Used by the Forest Products Laboratory at Madison, Wisconsin

COURSE 3:
STRUCTURE AND PHYSICAL PROPERTIES OF WOOD AND PLANTS

Professor Rogers

DESCRIPTION: The source, structure, and important properties of the raw, fibrous materials used in pulp making are discussed. Wood identification from a practical standpoint and suitability of the various woods for pulp purposes are included.

HOURS: 3 hours lecture a week.

COURSE 4:
MICROSCOPY OF PAPER FIBERS

Dr. Otto Kress

DESCRIPTION: The staining and identification of the various fibers used in papermaking and the analysis of fibre content of different kinds of paper are explained and practiced.

HOURS: 2 hours lecture a week.

REQUIRED TEXTS:

- *TAPPI Paper Testing Methods*

COURSE 5:
PAPER DYING

Dr. Otto Kress

DESCRIPTION: The dyestuffs and pigments classified according to the method used in the industry are discussed from the standpoint of suitability, strength, fastness, etc.. Dyeings will be tested from a strength and money value.

HOURS: 1 hour lecture a week, 2 afternoons laboratory a week.

REQUIRED TEXTS:

- *The Manufacture of Pulp and Paper*
  Volume IV

COURSE 6:
PAPER TESTING

Dr. Otto Kress

DESCRIPTION: The chemical and physical methods used in paper testing are discussed and practiced.

HOURS: 2 hours lecture a week

REQUIRED TEXTS:

- *The Manufacture of Pulp and Paper*
  Volume V
- *TAPPI Paper Testing Methods*

COURSE 7:
PULPING PROCESSES

Dr. Otto Kress

DESCRIPTION: The mechanical, sulphite, soda, and sulphate processes, as well as the newer processes such as the Semi-Chemical and Keora Processes, are discussed. Visits will be made to the mills adjacent to Appleton where these processes can been seen and studied in operation.

HOURS: 4 hours per week.

REQUIRED TEXTS:

- *The Manufacture of Pulp and Paper*
  Volume III

COURSE 8:
TECHNICAL GERMAN

Professor Cast

DESCRIPTION: Advanced technical German to train the student so that he can use to advantage the technical German literature.

HOURS: 3 hours lecture a week.

COURSE 9:
ABSTRACTS AND BIBLIOGRAPHIES
Miss Roseth
DESCRIPTION: The methods and purposes of abstracting technical articles, the filing of abstracts, the methods of constructing technical bibliographies, and their use.
HOURS: 1 hour lecture a week.

In addition to the classes listed above, each student will be required to participate in a series of Outside Lectures during which experts on various subjects from the mills as well as representatives from manufacturers selling supplies and equipment to the industry will give talks on their special subjects.

When you get out of there with a doctor's degree it carries, shall we say, a lot of authority with it. When you get out and about in a mill and they find out you've got a doctor's degree in papermaking, what you've got to say is very definitely listened to.

— **Kenneth Hay**, entering class 1937

Clarence J. West

Clarence J. West born in Brighton, Michigan, in 1886. After earning a teaching degree from Michigan State Normal College in Ypsilanti, Michigan in 1908, West received his doctorate in physiological organic chemistry at the University of Michigan in 1912 under one of the world's leading organic chemists, Dr. Moses Gomberg.

West became interested in paper industry while working at Arthur D. Little, Inc., in Boston. He went to Washington, D.C. in 1921 to work as the director of the Research Information Service and editor of publications at the National Research Council, a position he held until he came to Appleton in July of 1936 as the editor of the Institute of Paper Chemistry.

West maintained his editorial position at the Institute until his death in 1952. His accomplishments and honors during that sixteen-year span are numerous. He was held in the highest regard by his students and colleagues alike. Dr. Louis Wise, an Institute research associate and professor, once said of West: "He was a man of strong opinions, and he would not tolerate sloth or shoddiness. He drove himself without mercy and he was admittedly an idealist." West's publications list at the time of his death numbered more than seven hundred books, articles, translations, and book reviews in the eld of pulp and paper technology. He was very involved in TAPPI, serving for more than thirty years as the chairman of the Committee on Abstracts and Bibliography. TAPPI recognized West's contributions to the organization and to the industry by honoring him with their Gold Medal in 1937. His acceptance speech aptly sums up his career at the Institute:

> "Since I have been at he Institute of Paper Chemistry, I have learned that, as far as I am personally concerned, I have been building, during the past fteen years, a reserve on information which enables me to answer, with a minimum of time and eort, the questions thrown at me by the sta of the Institute. And knowing those men, I feel sure that you will acknowledge that they are ace high as pitchers when it comes to throwing questions."

I had good rapport with the library personnel, including Dr. West. He was a stamp collector and shared with Berwyn Thomas and me. I recall how grateful I was to have him give me an early U.S. stamp titled "Cattle in a Snowstorm." I cannot recall the denomination but it was black and dated 1898.

— Bob Miller, entering class 1936

Outside the Classroom

We'd often go downtown and drink beer.
— **Al Schenck,** entering class 1938

One Saturday, Johnny Bard and I took a canoe trip up the Fox River, putting in behind Sage Hall. Up river we made portage over the dam and onto Lake Winnebago. After paddling a considerable distance down the lake, some weather set in with an adverse wind and rain. The waves came up and we were not in the safest circumstances. We beat a hasty retreat back to the dam. It was an extremely grueling task fighting both wind and waves with some bailing to boot. We were rather shot when we reached the river again. I do not believe I have ever worked so hard in my life.
— **Bob Miller,** entering class 1936

We could fit in with anything that was going on at Lawrence because we had student passes. We had typical college fun. Of course, we dated the Lawrence girls. They were usually very anxious to date us, the older guys. We went to football games and went skating.
— **Kenneth Hay,** entering class 1937

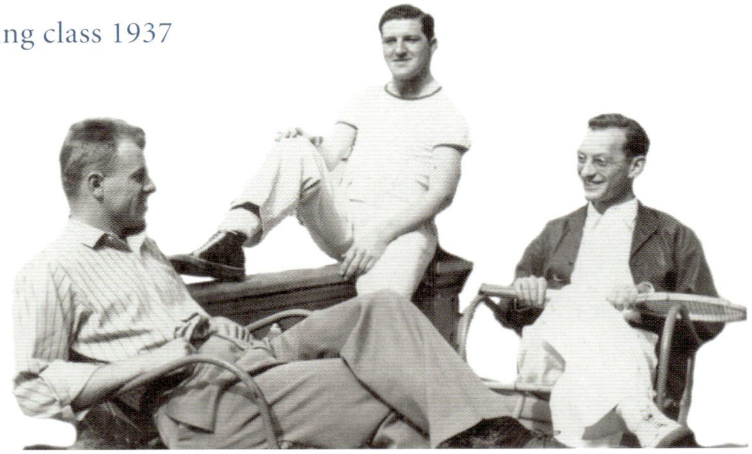

It came to the point that we didn't like the food that was served. We convinced Harry Lewis to give us the equivalent money to buy our own food and started cooking for ourselves. We had a hot plate and would make simple things like Campbell's beans and Campbell's soup. We even made toast on the hot plate.
— **George Ingle,** entering class 1938

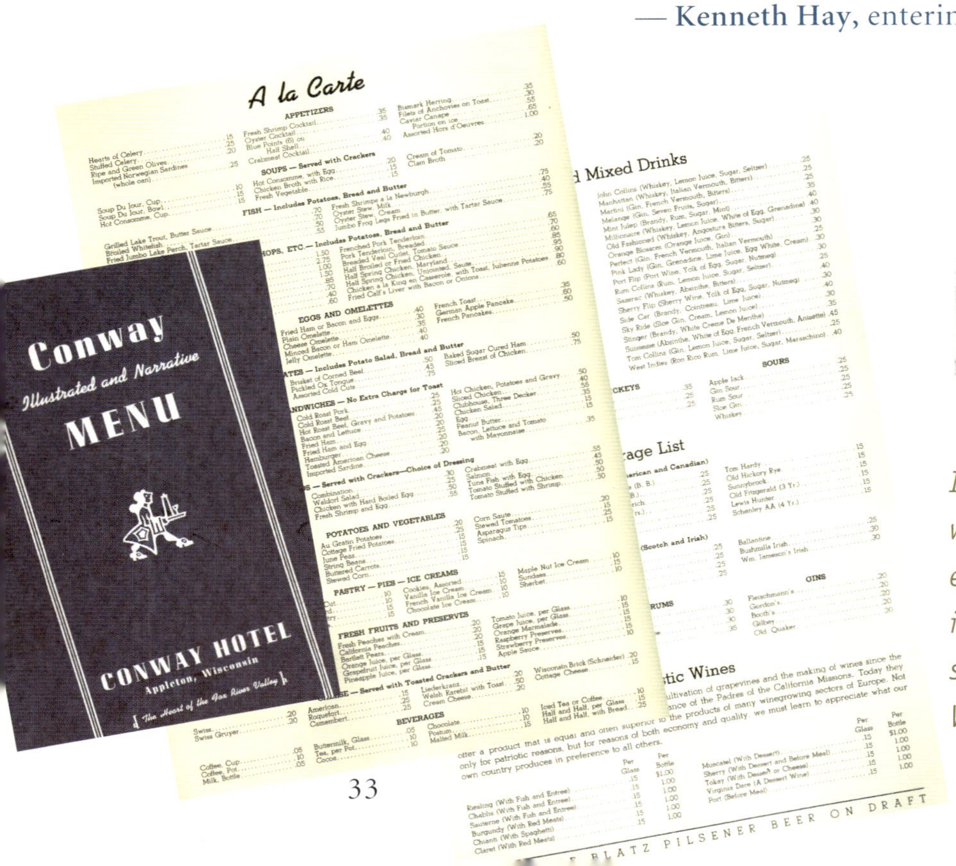

In East House, where we lived with three other couples, we got into a long-range bridge contest with another couple. The boys against the girls – if the girls won, we would have to take them out to dinner. If the boys won, we were gonna have a big dinner cooked by our wives. I don't know how many months we played this bridge game – we just kept adding to the scores. After several months, we decided to try and figure out who won. We added all the scorecards and I think the girls won. What was amazing was that after all the hundreds of thousands of points, the difference was like one hundred and eighty or something like that. We still laugh about it.

— **John McEwen,** entering class 1939

We would have stag parties where we would all dress in tuxedoes or suits and go to restaurants in town. Everybody would always say, "who are all those guys all dressed up?" But we just did it to have a little fun dressing up.

— **Daniel Adams,** entering class 1939

We used to go to the basketball and football games as a big group. We played a lot of handball at the gym across the street. We sometimes played three-handed with three people. That was a lot of fun and a lot of exercise.

— **Edwin Loutzenheiser,** entering class 1939

The Bush Trip in the 1930s

One of the quintessential Institute experiences is the Bush Trip. Though it has evolved significantly over the years – from that first one-day hike into Wisconsin's North Woods in 1930 to the one-month course in wood technology taught by Dr. Isenberg at Trees for Tomorrow to the trips of today sporting more modern conveniences – the primary goal of the Bush Trip has not changed. It was, is, and hopefully always will be an opportunity for students to get to know the men – and, increasingly, the women – with whom they will be sharing all the ups and downs of their Institute experience.

We went north to some logging camp. It was pretty much directed toward pulp wood sticks and was quite complete in operation. For some reason, toward the end of the day, I became separated from the group and they left without me. Guess there was no head count. The logging camp treated me to a sumptuous dinner while waiting for a car to return to pick me up. That was my introduction to the Institute of Paper Chemistry.

— **Bob Miller**, entering class 1936

I have wonderful memories of that trip. Professor Isenberg was with us – he was a wood specialist. We visited a logging camp, we visited a paper mill, and we visited a seeding plant. As a class, we became well acquainted.

— **Al Schenck**, entering class 1938

We had a bus and Irv Isenberg, who we called coach - a very responsible guy, a very cheerful guy, and very informative - he was a leader. I forgot the name of the bus driver but he was a jolly soul. Isenberg made the trip very useful. It was a lot of fun and we learned a lot. One funny thing - I was the only one who brought swimming trunks. There was Golden Lake there so we decided to go for a swim. So I said anybody who wants to borrow my trunks can, but I go first. I have to put on cold, wet trunks. So I went first. It was cold, miserably cold. I didn't stay in long. Then Ken Arnold borrowed my trunks and jumped in, then jumped out immediately. He couldn't take it. That was a fun experience. We got to know each other - a lot of sociability.

— **George Ingle**, entering class 1938

The experience stood me in good stead. The Institute has certainly made a lasting contribution to the industry and can be proud of its record. I have no regrets.

— **Bob Miller**, entering class 1936

We would get up every morning and go in and have breakfast. There was a table full of makings of sandwiches for the day and then we go out and work with the lumberjacks. At first I saw these guys carrying these big bags and I thought 'I don't need a big bag of food' but after the second day I was carrying a great big bag. You'd be out there early in the morning and you'd need lunch and snacks. We used to jump in the lake to clean up with soap. One thing we found out was the lumberjacks did not like snakes. So you could have fun by capturing a few snakes and putting them in your shirt or around your waist and if they were giving you too much trouble, then you took out the snake. It was a very good way to become acquainted with the fifteen guys you were going to be living with and studying with for four years.

— **Edwin Loutzenheiser,** entering class 1939

CHAPTER 2

1940-1949

It was devastating. We came home for lunch and there was a bridge game going. We were all watching the game. Then the radio came on and said Pearl Harbor had been attacked.

— **Daniel Adams**, entering class 1939

On that fateful date - it was a Sunday - a bunch of the first year group had just returned from lunch at one of the restaurants along College Avenue and were together in one of the bedrooms at South House, where we lived the first year. We were listening to the radio and heard President Roosevelt broadcast that this was a day that will "live in infamy!" We all thought, at that point, that we would not be students attending the IPC very long but would be in one of the military services serving our country.

— **Ben Haug**, entering class 1941

The Institute had closed out its first decade with a bang. Paper companies representing all four corners of the United States were applying for membership at a rapid pace, cutting-edge equipment was being brought in daily, and some of the country's most brilliant scientific minds had come to teach at the Institute. More than three hundred applications were coming in each new school year, each begging one of the coveted fifteen spots open each fall. The future certainly looked bright.

In spite of the upbeat outlook on life in Appleton at the end of the 1930s, the Institute kept a wary eye on events unfolding across the Atlantic. American involvement in the war would be devastating to an institution built around young men of draft age…

The effects of World War II on life at the Institute of Paper Chemistry can hardly be underestimated. Rations and shortages caused by events in Europe and Asia strained the Institute's resources tremendously, and educational activities virtually ceased for more than a year as the vast majority of students left their studies behind to join a branch of the armed services. Research work at the Institute shifted gears as scientists offered their services to the United States government; by 1942, the Institute was carrying more than thirty projects directly related to national defense.

World War II continued to impact the Institute even after the cessation of hostilities. Most students returned to their studies older, wiser, more world-weary – and in many cases married, forcing the Institute to revise its policy on married students. As Walt Bublitz, entering class 1941, explained, "The first couple of years after the war were a mess. You had fellows who came trickling in over different classes and years. Trying to fit all the students together and get classes for everyone caused many headaches."

The Institute's second decade of existence began smoothly enough. The budget grew from a mere $30,000 in 1930 to more than $330,000 by 1940 as more than fifty paper companies became members of the Institute. Much of the extra money was used for student scholarships; by 1940, the Institute was offering yearly scholarships of $650 for first-year students, $700 for second-year students, $900 for third-year students, and $925 for fourth-year students. The staff, numbering just three in 1930, had grown to ninety-five in 1940 and the equipment used by Institute staff and students was worth more than $1.2 million – impressive when compared with the $50 worth of equipment used by the first entering class.

Irving Isenberg

Dr. Irving Isenberg was born in 1909 in Buffalo, New York. He attended Buffalo Technical High School and the College of Forestry at Syracuse University, where he received a bachelor of science degree in 1931 and a master of science degree in 1932. His PhD in forestry came in 1936 from the University of California-Berkeley. Isenberg joined the IPC staff in 1937 as a research associate and instructor in fiber microscopy.

In 1940, the Institute created the Wood Technology Group and named Isenberg its Group Leader. In addition to teaching courses in forestry, wood and fiber chemistry, and microscopy and running the fiber identification laboratory, Isenberg was also responsible for the tree plantation that was cultivated in the valley behind the Institute.

Isenberg spent his entire professional career as a member of the Institute's faculty. He was the author of numerous publications, including two books, *Pulp and Paper Microscopy* and *Pulpwoods of the United States and Canada* that are still used today. Isenberg retired from the Institute in July of 1973 and was named Professor Emeritus, a title which he held until his death in 2001.

Who plants the trees and measures their span?
Who is this weighty, solid man?
Both chemist he, and botanist
Compiler and microscopist.
Teacher and forester in one –
Woods he explores, by grain – or ton.

— *The Insti-Tooter*, September 1952

I still feel that they [Institute people] are some of the best people I have ever known.

— **Victor Burstein,** entering class 1940

The Institute's course offerings changed as well; a necessity, thought the school's administrators, in order to keep pace with a rapidly changing industry. A Wood Technology Section was created under the direction of Dr. Irving Isenberg and, in connection with this group, a tree plantation was cultivated in the valley behind the Institute's main buildings. A new course for second-year and fourth-year students was added – speech coaching, taught by Lawrence's Ed Schoenberger, who later became the IPC Dean of Students. This class was initiated in conjunction with a new requirement that all PhD candidates must present a talk on their research work before the entire faculty. Schoenberger instructed IPC students in the art of speaking distinctly, forcefully, and with absolute poise. "The ability to think on their feet," Schoenberger wrote in an Institute newsletter detailing the new course offering, "will stand the boys in good stead after they leave the Institute." The course proved very successful; by 1941, it had expanded into a four-year program.

The early 1940s saw a flurry of construction and the arrival of cutting-edge equipment. In June of 1941, the Institute received its first electron microscope. The microscope – only the second one of its kind made by the RCA Manufacturing Company – was capable of magnifications up to 30,000 diameters and through further photographic enlargements, up to 100,000 diameters, allowing its users to view particles as small as 1/100 millionths of an inch.

In December of 1941, work began on a three-story addition to one of the wings of the Institute's present structures. The addition, completed in the summer of 1942, housed Dr. John Graff and his fiber microscope as well as the rest of the analytical and microanalytical, fiber microscopy, and wood technology sections, on the top floor; a semi-commercial lab, with storage space for supplies on the second floor; and new offices and labs for the engineering and physics sections, a room for the electron microscope, calibration facilities for the brightness tester, a first aid room, a dark room, a humidity room, two laboratory cubicles, and the spectroscopy lab on the ground floor.

Emil Heuser

In January of 1938, the subject of cellulose chemistry was forever changed at the Institute of Paper Chemistry with the appointment of Dr. Emil Heuser to the Cellulose Group. Bringing to the job, according to Westbrook Steele, an "unusual combination of theoretical and practical knowledge, both of cellulose chemistry and pulp and paper making, as well as rayon and its manufacture, gained from years of study and teaching and also from actual contact with these industries in Germany and Canada," Heuser proved an invaluable member of the Institute's staff during his ten-year stint as Group Leader.

Born in Stralsund, Germany, in 1882, Heuser studied at the Technical Universities of Munich and Karlsruhe under such famed chemists as Bunte, Engler, Haber, and Le Blanc. After completing his thesis at Austria's University of Graz, Heuser returned to Karlsruhe to obtain his doctoral degree. His dissertation, an experimental study of certain complex organic compounds supervised by the legendary Dr. Friedric Shcholl, was very well received in the field.

Upon graduation from Karlsruhe, Heuser spent three years gaining practical experience in various mills throughout Germany and Austria. In 1912, he became the chair of cellulose chemistry at the Technical University of Darmstadt, a position made vacant by Professor C.G. Schwalbe, then the leading exponent of cellulose chemistry in Germany. During his tenure at Darmstadt, Heuser's lectures on organic chemistry, chemistry of cellulose, and the chemical technology of pulp, paper, dyes, and textiles were often supplemented by his many original investigations on both theoretical and practical sides of cellulose, lignin and non-cellulose carbohydrates, as well as pulping, bleaching, and other issues pertaining to the manufacture of paper. In 1916, Heuser took a short leave of absence from Darmstadt to engage in a study of the industrial utilization of cellulose and other waste products. He returned to Darmstadt as a full professor in 1918 and remained there until 1923, when he resigned to take a position as the Director of Research at the Vereinigte Glanzsto-Fabriken, manufacturers of rayon and photographic films. In 1926, as the manufacture of rayon pulp became a pressing problem in North America, Heuser accepted a position as Director of Research of the Canadian International Paper Company in Hawkesbury, Ontario, Canada. Heuser held this position until his appointment to the staff of the Institute of Paper Chemistry in 1938.

Heuser is perhaps best remembered in the academic world for his lengthy bibliography of scholarly books and articles. His Lehrbuch der Cellulose-Chemie went through three separate printings in German, with the second edition being translated into both English and Russian. He edited six volumes of the Technik und Praxis der Papierfabrikation, as well as Cellulosechemie, which combined with Der Papier-Fabrikant in 1937. Heuser's most famous work, The Chemistry of Cellulose, a six-hundred-page tome still considered a classic in the field of cellulose chemistry, was published in 1944.

As a professor, Heuser was noted for his meticulous habits and old-school style of teaching. It was common knowledge among Institute men that should you happen upon Dr. Heuser's office before class you would inevitably catch him practicing the day's lecture, for his exacting nature demanded that each lecture be precisely fifty-five minutes long. George Martin, entering class 1941, remembers Heuser's reputation as a ladies' man. "He would come to all the dances and things in white tie and tails," Martin recollects. "You'd be dancing with your date and you'd see her smile and he'd be bowing to her. Next thing you'd know, he'd be tapping you on the shoulder to see if he could dance with your date. He was quite a guy."

Heuser was a fixture at the Institute until 1947, when health issues forced him into a lifestyle much more relaxed than the rigorous one offered by academia. During Heuser's farewell reception, Westbrook Steele aptly summed up the German scientist's career:

When Dr. Heuser came to us, it was his intention and his hope that he would find with us the opportunity to round out his career. Often he told me that in the Institute, he found the ideal opportunity for the utilization of his talents.

By August of 1941, the United States had, according to historian James M. Burns, "crossed the threshold from peace to war… as the war widened in Europe, the Atlantic lifeline came under attack, the British sought more aid, and public opinion changed to favor more intervention." Concern over the possibility of active involvement in the war led the Selective Service System to begin assigning classifications to all men of draft age. At IPC, all students in good standing who were not Reserve Officers were placed under Class II-A, the general action for satisfactory students of graduate technical schools. An investigation of the Institute's academic program and activities in May of 1943 led the Selective Service System to determine that the classification could continue, as, according to a letter sent to the school from the Field Service of National Headquarters, "the courses of study offered are preparing students for activities essential to the support of the war effort; that serious consideration should be given to the occupational classification of those students in preparation or training for whom you are willing to certify may be reasonably expected to successfully complete the courses of instruction in which they are engaged."

The months following the attack on Pearl Harbor saw a drastic alteration of the Institute's academic program. Citing the urgent need for accelerating the school's coursework, third-year IPC students remained in Appleton engaging in PhD research rather than leaving for the traditional summer job in the paper industry. This policy, which resulted in the graduating classes of 1943, 1944, and 1945 only spending two summers in industry, was to remain in effect through the duration of the war. Despite such hardships, the Institute maintained a positive outlook for the future. As Westbrook Steele told the assembled industry fellows at the 1942 Executives' Conference, "The Institute started in a critical period; it has grown through a critical period in the belief that it would be available for the crises that lay ahead and that its performance would justify its continued growth through the results it achieved for those who maintain it."

Dr. Al Howells

After I graduated from the Institute and worked for Camp Manufacturing Company, later Union Camp Corporation, I usually attended the annual TAPPI meeting in New York City and went to the IPC alumni dinner. About 150 alumni attended. Harry Lewis would introduce each of the alumni telling where they worked, what their job was, and some news about the family of the alumnus. His performance reinforced the feeling that the Institute and its alumni were all one family.

— **Keith Bradway**, entering class 1948

The Institute's contributions to the war effort were enormous. As Harry Lewis wrote in the July 1942 newsletter, "suffice it to say that when peace is declared, we will have made our contributions." Within six months of the United States' entrance into the war, the Institute was involved in more then thirty national defense projects. Among these were the development of new antiseptics for use in the paper industry and other fields; continuation of the wetstrength gum investigation; discovery of new substitute materials for those no longer available; camouflage of paper, textiles and other materials; and the development of a new source of tricresyl phosphate and paracresol, which proved of great interested to the United States Navy. Two former Institute students were involved in the technical development of synthetic rubber, while another was instrumental in the creation of a high active gas. Dr. Al Howells, chief of the Institute's Plastics Section, worked closely with the Vidal Research Corporation on the use of nonphenolic resins for paper-based plastics, while Dr. Boris Berkan supervised research work on the utilization of milkweed by-products.

The most famous of the wartime undertakings was, of course, the Paper House Project, developed by Institute scientists under contract with the Office of Production Research and Development of the War Production Board. Erected in August of 1944 and made entirely from waste paper pressed into panels, the Paper House was designed to meet a need for temporary housing for people left homeless in war areas – "a low-cost structure able to resist tropical heat or arctic snow, rain or wind."

The Institute's First International Student

When the IPC Board of Trustees met in October of 1929, contemplating requirements for admission, they decided unanimously that only American citizens would be considered. Dr. Kress, in a 1932 speech to a group of paper mill superintendents, said the decision was based on the sheer folly – "ridiculousness" Kress called it – of spending money supplied by the American pulp and paper industry to educate foreign men who would very likely return to their home countries and support their own pulp and paper industries whose "competition is absolutely ruinous to the paper industry in our own country." So, despite the dozens of applications received from interested students from Japan, China and India, the Institute stuck firmly to its policy and admitted only American citizens.

Just as World War II changed so many of the Institute's original policies, it also changed the status of international applicants. In August of 1947, the Institute's first international student, Biswanath Sen – dubbed "Ben Sen" by his classmates - made the trip from his home in Calcutta to begin classes. Sen's five-year stint at the Institute, culminating in a PhD in 1952, was quite successful; after his first full year at IPC, he was honored with the Paper and Twine Associated award for outstanding performance.

Keith Bradway, entering class 1948, recalls one amusing encounter with Sen during the 1948 Bush Trip

I found some of the features [of wood technology] fascinating. I would sometimes exclaim at what I saw with expressions like "Wow!" or "Holy Cow!" Ben Sen, who was from India sat at the table with me and after a few days told me with great mock sincerity, "Keith, I don't say things like "Jesus Christ!" I don't think you should say "Holy Cow!"

The Paper House came in a bundle of materials that weighed 1,029 pounds and could be assembled in about an hour. The paper panels, which were treated to be weather and fire resistant, were strong enough to withstand a load of more than 6,000 pounds. The house needed no framing or studding; rather the panels were joined by a system of metal connectors that did not show from either side of the panels.

The Paper House was designed to accommodate four people. The structure's designers, led by Dr. Johannes Van den Akker, furnished the house with double-decker bunks to show how shelter could be provided for a family of four. They even fashioned some paper furnishings, including cellulose windows and a paper window shade.

The Paper House stood on the IPC grounds until 1952, when it was dismantled. During its eight-year existence, the structure withstood hot weather, extended periods of rain and mist, winds reaching speeds of more than sixty miles per hour, and several blizzards. Unfortunately, the house was never called into military service. The British government showed strong interest in the structures in late 1944, but their request for 250,000 of them proved too great for the chipboard scarcity brought about by the war.

By May of 1943, seven IPC students had asked for a leave of absence and volunteered for service, while six others, already reserve officers, were immediately called to duty. That left only seventeen students to make up the second-, third-, and fourth-year classes. Many staff members requested leaves as well, while others abandoned their regular research work in order to focus on new and vital paper developments for the war effort.

The last wartime degrees were conferred in May of 1944. The 1944-45 school year saw no new students matriculating; the May 1945 graduation ceremonies represented the first time that Lawrence College had awarded no IPC degrees since 1931. Five students entered for the 1945-46 school year. Two left for the armed services within one term and two more were unable to keep up with their coursework and did not return for a second term, leaving just one regular student in the class. The 1945 Executives' Conference was cancelled due to wartime travel restrictions; it was the first year since 1936 that no conference was held.

Friedrich "Fritz" Brauns

A native of Germany, Brauns pursued his education at several prestigious universities, finally settling at the University of Berlin, where he studied under the Nobel Prize-winning organic chemist Emil Fischer. After earning his PhD in 1914, Brauns continued to gild his own organic chemistry resume with work at the University of Munich, Waldhof Cellulose Works, and then at the Department of Cellulose and Industrial Chemistry at McGill University in Montreal.

Brauns was appointed to the staff of the Institute of Paper Chemistry in 1935 as a research associate and instructor in organic chemistry. It was here that he put his well-deserved reputation in the field to the test. To date, no one had been able to isolate lignin without the use of strong reagents, which altered wood from its natural state and as a result, destroyed its use in papermaking. To isolate lignin would provide the industry with an important source of income from what had traditionally been a waste product.

It took Brauns some time, but he eventually succeeded where no one else had, forever leaving his mark on organic chemistry and the pulp and paper industry.

Brauns, known as "Mr. Ligneer" to his students, retired from the Institute in 1955. As Professor Emeritus, Brauns continued as a consulting chemist for several years. In 1960, he published a supplement to his 1952 magnum opus *The Chemistry of Lignin*, noting advancements in the field from 1949-1958.

Dr. Brauns I remember as a character with a heavy accent. During one preparation in an open hood in his lab, I recall smoking a cigar as an early warning of the evolution of a noxious gas – the cigar would take on a foul taste in its presence.

— Bob Miller, entering class 1936

Fritz Brauns - he was a real German immigrant type and very gruff and hard to get to know, but beneath it all he was a lignin chemist.

— Walt Bublitz, entering class 1941

Fritz Brauns - he had flown airplanes for the German airforce in WWI. He was a meticulous character and he insisted that we had good laboratory procedure habits.

— George Martin, entering class 1941

Fritz Brauns and Buck Buchanan gave an oral test as a final for the lignin course. I did not do well on oral exams, growing flustered and feeling it necessary to give a quick answer. One question put to me was, "What color is given by phloroglucinol by angiosperm lignins?" That should be one of the simplest and most basic questions about lignin. But one of the students had been experimenting with a bright green color developed when spruce wood was contacted with fuming hydrochloric acid. Flustered, I said green. Fritz turned to Buck and said, "I thought that if I demonstrated the red color of the phloroglucinol test in class they would be sure to remember it. But I was wrong!" In spite of that I passed the course.

— Keith Redway, entering class 1948

George R. Sears

A physicist with an extensive background in chemistry, Dr. Sears was at home in all phases of our academic and research activities. His intellectual curiosity was unlimited, and over the years he gained an understanding of the technology of this industry that very few have ever rivaled. A teacher by instinct, he encouraged and stimulated both students and staff.

— John Strange, 1965 Executives' Conference

George Sears was born in Beaver Dam, Wisconsin, in 1904. After a short stint at Ripon College, he transferred to the University of Wisconsin, where he received both his B.A. and M.A. degrees. He spent four years as a chemist at the Du Pont Experimental Station in Wilmington, Delaware, before returning to the University of Wisconsin to complete his PhD in 1936.

Sears came to the Institute in late 1936 as a research assistant and a professor of physics. He was promoted to research associate in 1941 and transferred into the physical chemistry section in 1948, where he was named Group Leader. His final assignment, which he held until his death in April of 1965, came in 1953 when he was designated Chief of the newly formed Graphic Arts Section.

A short vignette published in the December 1949 issue of the *Institute Reporter* serves as an apt illustration of the character of Dr. Sears:

When paper men are in distress
Their troubles come to G.R.S.
Who with his active, loyal crew
Will see each "physchem" problem through.
George has no time to mope or nap
With all those headaches in his lap.
They tell me though that he's a menace
When on the courts playing tennis
Despite his silvery, thinning hair
He's here, and there, and everywhere.
With all his friendliness, I'd hate
To meet George in a hot debate.
His arguments have wider scope
Than an electron microscope.
His thrusts are keen, but I should mention
They just produce surface tension.
No words of his give rise to tears
He may be sharp – but never Sears.

I relish one exchange I had with George Sears who taught a course in electricity. He used the term bias, meaning a voltage deliberately applied on a vacuum tube. I asked him what it meant. George asked, "Do you do any sewing?"

I knew the use of the term bias in sewing, and I had done some, but I didn't want to admit it so I said no.

George said, "Well, if you don't do any sewing, why don't you do some reading?"

That joke at a student's expense was totally unlike George, but his analysis of my study habits was right on the money.

— Keith Bradway, entering class 1948

The cessation of hostilities in Europe and Japan were cause for celebration throughout the world. As the shouts of jubilation faded, millions of weary war veterans returned to civilian life. Many, anxious to reap the benefits of the GI Bill, chose to continue the education the war forced them to abandon. Others made the decision to pursue higher education for the first time. In the years immediately following World War II, enrollment at the nation's colleges and universities doubled, and university officials were faced with the seemingly endless task of finding classroom space and living quarters to accommodate the wave of new and returning students.

In order to avoid using capital funds for dormitories, the Institute had, in the past, rented space for student housing from Lawrence College. The post-war enrollment boom, however, ended this arrangement. Lawrence, facing its own housing crisis in 1946, reclaimed North House for its own use, leading to the Institute's purchase of a residence at 933 East College Avenue known as Mory House, intending to use it as a single student dormitory. In 1948, in celebration of Otto Kress's contributions to the creation and growth of the Institute, Mory House was renamed in his honor and was thereafter known as Kress House.

With the Institute's single student housing issues resolved, Institute administrators next turned to the problem

It has had a strong impact. I was never sorry I went there. It was a great career. The paper industry was a fine industry to work in. We were able to raise a family and get them all through college and grad school on my paycheck. It was an interesting time leading an interesting life.

— **George Martin,** entering class 1941

of accommodations for the married members of the student body. This had never been an issue, for until World War II, only unmarried students were allowed to matriculate. By 1943, it was obvious that the Institute was going to have to revise this policy; in his August 1943 newsletter, Lewis reiterated a comment made by Ted Danziger, entering class 1942: "By the time we all get back to the Institute we are going to have to worry about what to do with our children in addition to Dr. Van den Akker's mechanics course."

Some married couples made their homes in East House. The Alton Street residence could accommodate four married couples in two-room apartments with a common kitchen area and a study area set up in the basement. Other couples lived in West House (formerly called the A.B. Fisher House), which was designed in a manner similar to East House and housed six married couples. Finally, in

Sidney D. Wells

Sidney D. Wells was born in Boston in 1885. He graduated from New Hampshire's famous Phillips Exeter Academy in 1903 and received his Bachelor of Science degree from the Massachusetts Institute of Technology in 1907. After a few years as a chemical engineer for the Mead Pulp and Paper Company, Wells joined the staff of the United States Forest Products Laboratory at Madison as an engineer in 1910. He remained there until 1925, working on a variety of research projects, including the preparation of *Bulletin 1485*, a summary of pulping and grinding characteristics of a large variety of American wood for the United States Department of Agriculture, and a series of studies on the value of flax, straw, and cotton linters as papermaking materials.

Wells left the Forest Products Laboratory in 1925 to become the director of the Paper Mills Laboratories, Inc., in Illinois. In 1930, he went to work as the technical director of the Nekoosa-Edwards Paper Company, where he developed a successful multi-bleaching technique for kraft pulp. In 1932, Wells became the technical adviser for the Combined Locks Paper Company, perfecting a method of de-inking groundwood paper, working particularly with old telephone directories. He joined the consulting firm of Paulson and Wells in Appleton in 1935, where he remained until he came to the Institute as a research associate and an instructor in pulp and paper technology in 1938.

Wells' fifteen-year career in academia was as distinguished as his experience in research and in the paper industry. He published more than thirty papers on a wide variety of subjects ranging from the effects of adhesives on fiberboard manufacture, the use of sodium sulfite in alkaline digesters, the utilization of spent liquors, and low sulfidity kraft pulping to the cooking of bagasse, straw, cornstalks and cottonwood.

Wells retired from the Institute in August of 1953 and was given the title of Research Associate Emeritus. At Wells' retirement celebration, his friend and colleague Dr. Louis Wise honored him with the following tribute: "His achievements, and they were many, depended on thought and on quiet, painstaking effort."

1946, the Institute acquired eight Quonset Huts, which were immediately shipped to Appleton from Tacoma, Washington. These prefabricated, semi-cylindrical steel structures – "tin cans," as they were commonly called by soldiers and students alike – had been designed at the Naval Air Station at Quonset Point, Rhode Island, and proved an interesting solution to the housing shortages on college and university campuses across the country. The Institute's Quonsets were set up in a colony called West Court, where they provided fifteen apartments with a common laundry room. Under the original Lanham Act, which relinquished the units formerly owned by the federal government to private institutions, the Quonsets were to be removed by July 25, 1949; however, a new law passed in 1948 extended the date indefinitely. These unique living arrangements certainly changed life at the Institute; as Lewis pointed out in his April 1947 newsletter, "the general flavor of the Institute body has changed with the importation of all those wives and children."

Dr. Otto Kress

By the fall of 1946, life had begun to return to normal at the Institute of Paper Chemistry. The first "post-war" class arrived in Appleton on August 26, 1946, for their orientation, while many other returning veterans attempted to pick up their graduate studies where they left off in the early 1940s. Plans were made for refresher training programs to help reorient the GIs to academic life. Each veteran represented a unique situation; Lewis and his administrators were forced to handle each circumstance individually.

Several important curriculum changes coincided with the return of the students who had left their studies behind to join the armed services. The Institute pared down "tool" courses, along with moving all the physics courses to the first year. Dr. Rowland introduced a new colloid chemistry class that dealt with the physical properties of the high molecular compounds; he called it "colloid chemistry modernized." Additionally, the infamous third-year problems were introduced into the third-year program as a means of teaching how research is organized and determining whether the students had the capacity for planning and carrying on research.

It provided me with an opportunity, which, as time goes on, I appreciate more… it gave me an opportunity to do something that was difficult and it tested you – God knows it did. It taught you discipline. In my case, it taught me how to write and how to report and how to manage a project. I put it down as one of the truly significant parts of my life and my career. You never know what would have happened if I hadn't gone to the Institute but I think it gave me a status… having a doctorate and being in the paper business… and having gone an institute that's run and managed and paid for by the paper business… you got recognition that you wouldn't get otherwise.

— **Bill Haselton,** entering class 1949

The 1940s ended with a bang, although perhaps one of a little less magnitude than the one that began the decade. Otto Kress, the Institute's first full-time staff member, announced his retirement in February of 1948, closing what many view as the first chapter in the Institute's history. At the same time, the Institute began a more intense recruiting program to bolster flagging interest in graduate education, a trend seen in graduate schools throughout the country. Lewis opined that many World War II veterans felt that they were too old to continue their educations once they had completed four years of undergraduate training.

By the end of 1949, the Institute boasted more than two hundred staff members and a $2 million research center. The one hundred and seven member companies represented seventy percent of all pulp and paper production in the United States, while the Institute's library had grown to more than thirteen thousand volumes, making it one of the largest and most complete collections in the world. As John Strange pointed out at the decade's end, "the steady and virile growth of the Institute through the Depression and war years is an abundant verification of the soundness of the framework which was originally laid down by the charter members."

The Westbrook Steel Medal

In 1941, Westbrook Steele, one of the key figures in the development of the Institute of Paper Chemistry, established an award in his name to be presented to a graduating PhD student. The Westbrook Steele Medal was to be awarded to a student who, through his or her dissertation, possessed an understanding of the relationship of the subject of investigation to the broader problems of pulp and papermaking; who demonstrated originality and soundness of approach to the subject; who showed ingenuity and thoroughness in his or her techniques; and who exemplified craftsmanship in preparation and presentation of the dissertation.

Westbrook Steele Medal Winners

1941..........Robert A. Stillings
1942..........William H. Lane
1943..........Daniel O. Adams
1944..........Charles M. Sigvardt
1947..........Arthur J. Haug
1948..........John S. Barton
1949..........Joseph C. Brown, Jr.
1950..........Ellsworth H. Shriver II
1951..........William L. Ingmanson
1952..........Thomas G. Zentner
1953..........William R. Haselton
1954..........Samuel M. Rollinson
1955..........Robert W. Hisey
1956..........Arthur C. Dreshfield, Jr
1957..........John R. Parkinson
1958..........William H. Wiseman
1959..........Charles W. Carroll
1960..........Richard H. Cornell
1961..........Ronald B. Estridge
1962..........Edwin W. Arnold
1963..........Frederick J. Vermillion, Jr.
1964..........Carl A. Gentzen
1965..........Philip H. Emery, Jr.
1966..........Harold L. Hintz
1967..........Robert J. Morgan
1968..........Charles E. Dunning
1969..........R. Paul Kibblewhite
1970..........Karl I. Ebeling

Life in a Quonset Hut

The post-World War II era brought into effect several changes in Institute policy and procedure. Among the more notable was the end of a rule that had been in effect since the Institute's inception – that a student must remain single in order to maintain his scholarship. Naturally, students found ways around this policy. Jean Hess, wife of Cecil Hess, entering class 1946, recalls the story of one student who had been repeatedly warned of his "fraternization" with a female during his time at the Institute. "After he received his diploma and was assured that it was signed," Mrs. Hess remembers, "the young man introduced his wife to the dean!"

The Institute was forced to review and revise this policy when the World War II classes returned to resume their studies. In addition to military experience, many of these young men also acquired brides during their time in the service. Desperate for housing for these married couples, the Institute acquired eight Quonset huts that were placed in two rows in a field along West Court Street.

What follows is a collection of recollections and remembrances of the experiences of several Institute students who lived in the Quonset Village. "During the bad times when the icebox water ran all over the floor – when we were cold because the oil had run out - when someone had not followed the wash hut schedule and we had diapers to wash or dry – when a cloudburst or blizzard had impeded our husbands' trek to the Institute labs – when new babies cried all night – when budgets were running low," laughs Mrs. Hess. "We would look at each other and say, 'I suppose some day we will look back on all this and laugh!' And now, we do."

They weren't the most luxurious of accommodations. I can remember how cold they were in the Appleton winters. You had a space heater in there, an oil-fired space heater, and you had an electric fan up in the dome to blow some of the warm air back down again… and we had honest-to-goodness iceboxes. An ice man came around with a cake of ice that you bought and put in there. You had to remember to put a pan under there for when the melted ice water dripped and you had to remember to empty that. After once forgetting to empty the pan, I came home from the lab with a drill and a funnel and made modification number one. I drilled a hole underneath the icebox and put the funnel in it and therefore never had to worry about emptying the thing again. Pretty soon, every other Quonset hut had adopted my procedure.

— George Martin, entering class 1941

It was interesting… they were split in half and there was a married couple in each end. They consisted of three rooms and a bath and a shower. There was a living room/kitchen, and then one bedroom, and then another small bedroom/study, heated by a central oil-burning stove, very little insulation either in the walls, ceiling, or floor. It was cold. But it was survivable.

— Bill Haselton, entering class 1949

In the wintertime when it got really cold, the internal layer was held to the exterior metal layer by metal rivets and in the mornings they'd have ice on them… we shared our Quonset with a guy named Jim Schoettler and his wife. The way they were set up was that there was a fire escape that was a door about a foot tall and two feet wide between the two bedrooms in the back of each of the halves. We got to know the Schoettlers really well by visiting back and forth through that little door.

— George Mead, entering class 1950

We had a freshly painted one-half Quonset at 16 West Court. The paint job was a horrible pink. First order of business was to repaint. The color we selected – a pleasing brown - was viewed as extreme by the Institute's painter but it was popular at the time.

— John Parkinson entering class 1951

Living in a Quonset hut was comfortable enough except for the ice on the ceiling in the winter. You got to be so close to the person next to you… you got to know each other pretty well.

— Richard Barker, entering class 1958

The thing that both Mary Anne and I remember the most is living in the Quonset Huts. They were cold and they smelled… thank God we didn't have any kids back then!

— Joseph Koleske, entering class 1958

You could stand in one place and you could be in every room at one time. You could always tell how cold it was by how much frost was on the nails in the Quonset. Then in the winter, the rabbits would come underneath the Quonset huts. There was aluminum foil for insulation and you could hear them scrambling around down there.

— Carl Jentzen, entering class 1959

Outside the Classroom

My home was in Neenah and I would go home on weekends fairly often and I recall on Thanksgiving, my mother would ask me to bring the fellows home. My mother was quite delighted to see her turkey disappear.

— **Victor Burstein**, entering class 1940

The $650 scholarship provided South House boarding for 9 months. I did my laundry in the basement. It provided for three quarters of tuition, books, and lab fees. It provided three meals a day in Appleton restaurants for nine months. I ate three meals a day for $1.00 a day. We got a break buying meal tickets - $5.50 worth for $5.00. We ate breakfast and lunch at Hairless Herb's Horrible Hash House- the Checkerboard restaurant on College Avenue on Lawrence Campus.

— **Ben Haug**, entering class 1941

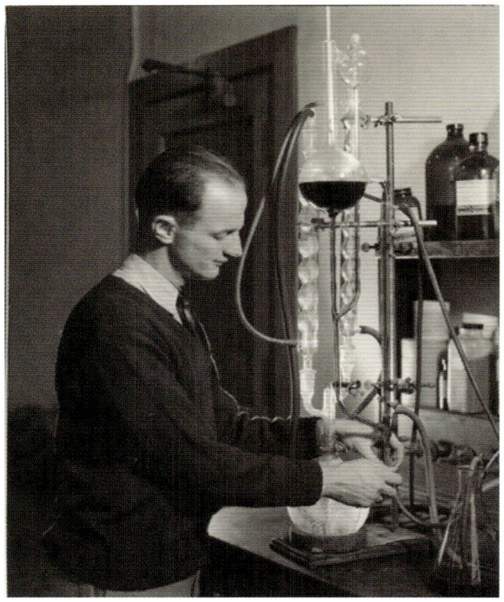

We liked to play volleyball quite a bit. We also like to play ping-pong and going to the football games.

— **Bob Boehm**, entering class 1948

Incidentally, in our group there were five of us that each pitched in $7 and for $35 we bought a 1926 Essex Super Six. We managed to keep that car going. One of the funniest things was when we went out to the Executives' Conference at the country club and all the executives were driving up in their fancy automobiles and they had valet parking. The five of us drive up in our Essex Super Six and we made quite a sensation. It lasted a long time.

— **George Martin**, entering class 1941

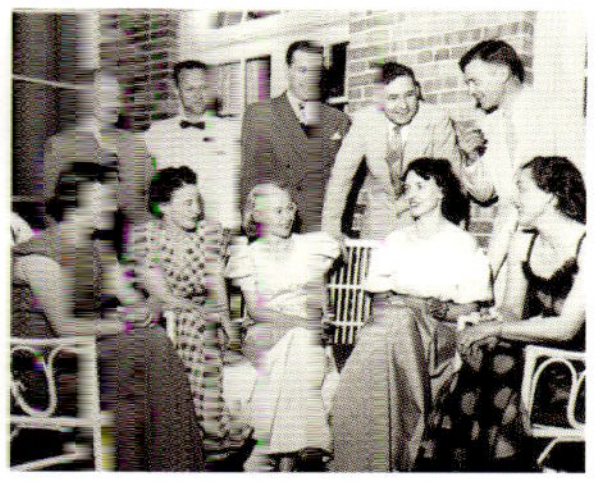

We had parties, usually in the dorm. This was when the Kress House was down the end of College Avenue. It was a big house that someone had given to the Institute. Bob McKee was on the faculty on the Institute for many years and he lived behind us and every time we had a party you could set your clock by it - at midnight, the police would be there at the door. We would wake him up every time!

— **Otto Kroeschell**, entering class 1949

We did a lot of partying... our class was inclined to party.

— **Dan Kane**, entering class 1948

A highlight of the year at Kress House was a homecoming party on the weekend of the Lawrence College Homecoming. As a part of the party a skit was given. In my second year I worked with Otto Kroeschell and Bill Hoge to produce a skit. We used mostly music from "South Pacific." The skit was called "South River" after the street location of the Institute. The song "There is Nothing Like a Dame" was converted to "There is Nothing Like a Brain." The song "Bloody Mary" became "Bloody Harry," a jab at Harry Lewis.

— **Keith Bradway**, entering class 1948

The Bush Trip in the 1940s

The Bush Trip – the traditional sojourn into the Northern woods that begins a student's IPC experience – received a makeover by the end of the 1940s. Since the trip's inception in September of 1930, the Institute had taken first-year students to the University of Michigan's forestry camp at Golden Lake. These trips ranged from one day to two weeks in the woods. In 1948, however, the trip was moved from Golden Lake to the Trees for Tomorrow camp at Eagle River and became a month-long course in wood technology taught by Dr. Irving Isenberg. At Trees for Tomorrow, the students were able to cover the subject by both precept and practice. In addition to classroom-style lectures, students visited the forest experiment station, forest rangers' station, the Hugo Sauer nursery, Rhinelander Paper Company, and the Goodman Lumber Company.

The first thing we did - we went to a lumber camp in northern Wisconsin. We were there for three or four days and learned to identify trees. I remember waking up hungry for breakfast and having pancakes that were very, very good.

— **Victor Burstein**, entering class 1940

It was two weeks in upper Michigan, along the shores of Lake Superior. It was an introduction to forestry, forest management, and ultimately pulp and paper. We stayed in the University of Michigan forestry camp. We went around to the different forestry operations up there, saw logging and hauling of logs into the mills. They had steam engines in those days that carried the logs in. To me it was absolutely fabulous because I had never been in the North Woods before. We had a great time getting to know each other. It was marvelous.

— **Walt Bublitz**, entering class 1941

It was fun; it was new to all of us because we were green.

— **Richard Brown,** entering class 1946

We visited several nearby paper mills and a sawmill, but Irv Isenberg found it hard to find modern woodcutting operations for us to watch. One day he was directed to a site where a lone worker was cutting down an aspen tree with a large bow saw. He was not working however, but sitting on a stump. One of the students asked Irv to see why he wasn't working. Irv dutifully approached the man and came back, saying, "He's just resting. He's tired from all that sawing."

— **Keith Bradway,** entering class 1948

That was a very cold summer up in Eagle River. We were up there a couple of weeks. It was interesting… we went to a sawmill, and a couple of paper mills… we went with the forest rangers out to look at the deer and the deer damage. It was to give you a sense of the forestry side of things because you didn't do much of that later. Later on it was more of the academic, scientific, chemistry, physics, and all the rest.

— **Bill Haselton,** entering class 1949

It was quite wonderful. We met in Appleton - 14 guys from all over the country. We were transported in a special Appleton bus. The bus driver was Andy Kangus. He was awfully nice. We went to upper Michigan. It was held at the University of Michigan forestry camp at Golden Lake, Michigan. We had a wonderful two weeks and got acquainted with one another and nobody shaved for two weeks. We looked motley by the end of the two-week period. We would play touch football games. It was getting acquainted with one another and soaking up tree knowledge. We went into Iron Mountain, Michigan, on Saturday night and we found a few places to relax and have fun. We had a good Saturday evening. By the time the two weeks were over, we were very well acquainted. We were all men at that time - no women students yet.

— **George Martin,** entering class 1941

I remember clearly being in a laboratory and looking at pieces of wood and trying to identify them. It was late at night and we went on field trips in the daytime and we had to identify wood by looking at it under a microscope. We did that at night after we got back from the field trips. We'd chat and holler at each other.

— **Otto Kroeschell,** entering class 1949

CHAPTER 3

1950-1959

The 1950s – which many historians agree actually began in 1946 with the end ofWorld War II and the return of hundreds of thousands of veterans to civilian life – was a decade symbolized by beginnings. Emerging from economic depression and shortages resulting from the war, Americans were eager to embrace the prosperity that seemed inevitable. The "baby boom" that followed the war increased the population of the country several times over; by 1954, annual births topped four million for the first time in history, a trend that would continue until 1965. A direct result of his population explosion was the birth of the American Dream, the desire of young American to live in relative comfort and stability in the suburbs sprouting up around major cities.

The Institute was not immune to these trends. By the early 1950s, more than half of the student body was married, and many had children. When the Quonset huts that had served as married student housing since the war's end became increasingly harder to maintain, the Institute was forced to reconsider its housing options. What followed was the Institute's version of the American Dream – its own suburbia in the form of West Court...

The 1950s was a decade of "booms." The baby boom, the industry boom, the building boom – the United States was alive and well and wanted the world to know it. The Institute of Paper Chemistry was no exception.

ON THE OCCASION of the Institute's Silver Anniversary in 1954, Dean Harry Lewis remarked that it was a wonder students could hear the lectures of their professors over the noise of the construction that was taking place. Indeed, construction was a key theme characterizing the Institute of the 1950s.

In 1929, the Institute's founders agreed to set aside $410,000 to build a complete modern semi-plant to be used for research and training. At the same time, however, The Royal Technical Institute of Stockholm announced its plans to create a similar plant. The Institute's founders agreed to delay their own project until the Swedish venture could be evaluated. When finally put into operation in 1949, it was deemed too costly for the Institute to maintain and operate to justify its use as a research and demonstration tool.

With the plans for the semi-plant cast aside, it became obvious that the Institute still required additional laboratory equipment and devices, which, in turn, called for additional space and facilities. The project was costly, as the designs called for heavy and expensive per-square-foot loads. A better option, it was decided at the 1951 Executives' Conference, would be to erect a new building of lighter, less costly construction, move some of the Institute's existing activities there, and then adapt the vacated space where floor load, electrical, and plumbing facilities were already available.

The Institute's building boom kicked off in the fall of 1951 with the renovation of the old semi-commercial lab. When the building was first erected more than twenty years before, several clips were left in the wall to potentially accommodate another story. By December of 1951, a beautiful mezzanine floor lay across the

J. Edward Todd

Dr. J. Edward Todd – "Ed" to his friends and colleagues, Dr. Todd to his students – was born in Vermillion, South Dakota, in 1897. After obtaining his A.B. degree from the University of Kansas in 1918, Todd enlisted in the Air Service, volunteering for Near East Relief to help with reconstruction following the First World War. After spending a year in Constantinople, where he was responsible for the transportation of relief supplies to the interior sections of the Caucasus area, Todd spent several years in academia, first at the University of Vermont, then as the business officer of Cornell College in Iowa. He returned to school and completed his M.A. in 1934 and his PhD in 1940, both from the Teachers' College of Columbia University.

Todd came to Appleton in 1942 to serve as the Institute's Assistant Dean. His main interests lying in the fields of higher education administration, psychology, and philosophy, Todd was best known to students for his administration of a series of aptitude exams during the first days of the bush trip.

Todd spent a total of twenty-four years at the Institute. He was appointed Dean of Admissions in 1950, a position he held until his semi-retirement in 1962. Todd served the Institute in a consultant capacity for five additional years before officially retiring in 1967. He passed away in 1973.

Dean Todd was a character, but he was really on the side of the students.

— Arthur Brashfield, entering class 1951

beams, creating a new space for Dr. Roy Whitney's chemical engineering laboratory and a home for equipment that had previously been scattered all over the Institute. A special explosion-proof room was added in the building's southeast corner to house the wood extraction unit and its related equipment as well as a stainless steel reaction vessel for alkaline fusions. A new humidity room and machine shop completed the facilities.

Construction on the new General Activities Building, first discussed at the 1951 Executives' Conference, began in the summer of 1952. Designed to match the present main building in size and appearance, the new building contained all administrative offices, business and education offices, classrooms, an auditorium designed to seat two hundred people, and a museum – twenty-one thousand square feet spread over two stories, a basement, and a sub-basement.

The General Activities Building was completed in the late spring of 1953 at a cost of $347,500. During the June 4 cornerstone ceremony – held indoors due to a heavy thunderstorm – the Chairman of the Institute's Board of Trustees, Ernst Mahler, dedicated the building "to the abiding faith we share in the future."

The General Activities Building contributed to the growth of the Institute in a variety of ways. The new laboratory and office space led to the expansion of the research staff, which led to the expansion of the research program. Moreover, the additional space allowed more students to attend the Institute. Before the expansion, the Institute could support a maximum entering class of fifteen and student body of fifty-five. With the new facilities, an entering class of twenty and a student body of seventy-five could be accommodated.

The enlarged student body – and the wives and children that accompanied many of them – resulted in the Institute's next building frenzy. In November of 1954, the Institute announced plans to construct five new apartment buildings for married students. Each building – two two-story units and three one-story units – was designed to provide housing for four married couples with a living room, a dining alcove, a kitchen, a bedroom, a study/nursery, and a bathroom.

The five apartment buildings were completed in July of 1955 at a cost of $275,000. A sixth building – another one-story unit - was completed the following September. A seventh unit was built in January of 1957 and an eighth in the spring of 1958. The Institute's own little suburb was named West Court; more often, however, it was simply referred to as "Fertility Row!"

The Institute has been very important in my life and I'm glad I had the opportunity to go there. You could do anything you wanted to do with an Institute education.

— **Vincent Russo**, entering class 1952

It was an awesome place.

— **Frederick Vermillion**, entering class 1958

I think the Institute gave you a well-rounded technical background and also taught you a way to think.

— **Carl Jentzen**, entering class 1959

The building boom continued throughout the 1950s. At the 1956 Executives' Conference, a $735,000 expansion program was announced, promising to add close to thirty-seven thousand square feet to the Institute's present quarters. Construction on the new spaces – an addition to the Container Building, a new building to house pilot plant equipment, and the enclosure of the court around the present buildings – began in May of 1957.

The addition to the Container Building was completed in January of 1958. The basement was renovated to house the Container Division, while the main floor was expanded to meet the increasing requirements in paper evaluation and graphic arts. Two new humidity rooms – one large and one small - were added, as well as a large chemical laboratory for graphic arts and five new offices for the staff of the groups occupying the building.

Shortly afterward, another building – South Laboratory – was completed at the rear of the campus, adding sixteen thousand square feet of laboratory space. The first floor was leased to the Sulphite Pulp Manufacturers' Research League, who had designed it to meet their needs for research and pilot plant facilities and for office space and conference rooms.

Finally, in July of 1958, the final stage of the proposed expansion program – the Court Addition – was completed. Designed to join the west rear of the General Activities Building with the southwest wing of the Institute's main research lab, the addition formed an enclosed inner court that was to be maintained as a beautiful green space. Lewis called the space a memorial to the Institute's former librarian, Edith Stroschneider, who passed away in July of 1956, for "her tulip bulbs still grow and bloom under the windows of the library…where she spent her rare periods of rest and recreation."

Dr. Philip N. Joranson

The Court Addition added twenty-three thousand square feet of lab and office space. Students loved the two new air-conditioned classrooms, while the space vacated by the moves allowed for more lab room for student thesis research as well as for expansion by the analytical chemistry, chemical engineering, and biology sections.

The completion of the expansion program – which ended up costing the Institute some $950,000 – did not end the 1950s building boom, however. In November of 1958, the Institute received a $50,000 grant from the Louis Calder Foundation for a new lab to be known as the Lou Calder Plant Biochemistry Laboratory. The funds, donated by Louis Calder, the President and Director of the Perkins-Goodwin Company in New York City, allowed the purchase of ultracentrifuge and electrophoresis equipment as well as complete and modern facilities for freeze-drying biological and biochemical preparations. One year later, the Calder Foundation contributed another

I enjoyed it from the time I got there until the time I left. A lot of the guys didn't - they couldn't wait to finish up and get out…they had other things to do. But I really enjoyed doing it; it was one of the most fun things I've ever done. It provided a solid foundation for a career in the paper business. Going to the Institute was one of the smartest things I ever did.

— **John Schulz**, entering class 1955

The technical aspect wasn't as important as the discipline it took. That was discipline that I needed. I learned to think of things in an entirely different way.

— **John Vaughan**, entering class 1958

$30,000 gift to establish an updated electron microscopy and diffraction laboratory that would permit diffraction studies not possible with the Institute's twenty-year old electron microscope.

The Institute's new research facilities spawned a flurry of research activity throughout the 1950s. One project that proved of supreme importance to the field of medicine was research done on vanillin, which ultimately resulted in a cure for histoplasmosis – otherwise known as "Valley Fever," a uniformly fatal disease in children. Institute researchers worked for several years with vanillin, a compound derived from spent sulphite liquor, finally discovering a way to turn it into vanillic acid. From there, ethyl vanillate was created and proved to be an effective weapon against fungi, the cause of histoplasmosis. In 1952, the Institute turned its findings over to a research team at Vanderbilt University – and then offered to finance the clinical trials. "So far as we know," Westbrook Steele told the 1952 Executives' Conference, "this is the first time that one educational institution on its own account and from its own research reserves had financed research at another institution of higher learning."

Shu Tang "S.T." Han

Originally from Hangchow, China, S.T. Han joined the Institute staff in 1949 as a research associate and professor in the Chemical Engineering Department. He left for a brief period in the mid-1950s for a position at Beloit Iron Works, returning in August of 1958 to continue his academic and research responsibilities in the chemical engineering lab.

In 1964, Han left the Engineering and Technology Section to serve as a research associate in the Dean's office, where his responsibilities included the formulation and implementation of programs resulting from both internal and external examination of scientific trends and movements engaging the interest of the Institute. In 1966, Han was named chair of the Engineering and Technology Section, succeeding the late Bill Ingmanson. He held this position until his retirement in 1977, at which time he was awarded Emeritus status.

S.T. Han was a very brilliant chemical engineer and professor, yet his love was cowboy movies. He wanted guys with white hats and black hats, the good guys and the bad guys. When S.T. would go to the movies, it had to be a cowboy movie or he wouldn't stay.

— Vincent Russo, entering class 1952

S.T. was brilliant and always helpful. The recollection we all have of him is standing in the auditorium writing these long equations or the blackboard with his right hand and dragging an eraser across behind him with his left hand so that you could hardly find time to get it all down in your notes before it was erased.

— Robert Jones, entering class 1957

The professor that made an impression on everybody in my day was S.T. Han, who taught chemical engineering. He was enthusiastic and took great pleasure in challenging the student. Everybody who took his courses felt challenged but really enjoyed them.

— Jim Burke, entering class 1961

S.T. was a heavy smoker. One day he was up there in the front of the classroom, drawing equations on the board and solving a problem. He had a cigarette in one hand and he had the chalk in his other hand, and when he was done, he put the chalk in his mouth! But nobody laughed, because S.T. was a pretty strict, sensitive guy, and if you laughed you might've flunked the course!

— Al Jones, entering class 1962

S.T. Han would walk around the room and then suddenly stop and ask you a question. No one ever wanted to answer the questions, but he wouldn't leave until you answered correctly.

— Robert Morgan, entering class 1962

George W. Mead

George W. Mead was born in Milwaukee, Wisconsin, on October 11, 1927. After completing his secondary schooling in Connecticut, Mead attended Yale University, where he earned his Bachelor of Science degree in chemistry in 1950.

"Really, I went through my senior year with loose ends, wondering what I was going to do with myself," said Mead. "I think there was sort of an understanding that I would come back here to Wisconsin Rapids and work for our company, at that time called Consolidated Water Power and Paper Company, but I had no credentials, really, other than a degree in chemistry and that didn't seem to me like it was going to carry me very far."

It was around this time that Mead saw a bulletin board posting advertising the Institute. The notion of a school devoted to the study of pulp and paper technology intrigued him and he sent his application a few days later. The answer from the Institute was positive; Mead was accepted into the entering class of 1950 with a full scholarship. "It was," remembers Mead, "quite a thrill."

Mead's Institute experience began with a bush trip. "We got acclimated to forestry practices in a very limited but hands-on way. We visited a whole bunch of various wood consumers, a lumber mill, and several pulp and paper mills throughout the area. That whole trip was very worthwhile. Besides that, it gave us a chance to get together, the fifteen of us, and get to know each other and that was a really important factor too," Mead recalls.

Upon returning from his bush trip adventure, Mead, along with his wife and child, settled into his half of the Quonset hut he shared with classmate Jim Schoettler. "They were…spartan," said Mead, "and, frankly, that might even be too generous a word! It was snug in the physical sense. It was not snug in the warm sense."

Mead quickly became acclimated to the Institute course work and routine. He fondly recalls professors and administrators such as Johannes Van den Akker, Irv Isenberg, John Green, Roy Whitney, and Harry Lewis. "There's a story kicking around out there about me trying to dry out a solvent solution in one of John's (Green) ovens and blowing the oven up, which is true. I did it," admits Mead. He also recalls the toughness of Van den Akker's classes: "He was the filter. As you got halfway through his course he threw a tough test at you and you found out who were the sheep and who were the goats right off the bat. As far as I could tell, that test was the one key element that sorted the class out as to who was going to stay and who wasn't."

In May of 1954, the Institute began a new research program in Forest Tree Genetics. The program, led by Dr. Philip N. Joranson and financed by Rhinelander Paper Company and Marathon Corporation, sought to determine how and to what degree genetic make-up influenced both qualitative and quantitative aspects of the growth of trees. The program's eventual goal was to breed a new, improved species of tree that would yield more tons of paper per acre of forest.

By May of 1955, Joranson's research into the genetic make-up of trees uncovered a variety of species with more than the normal two sets of chromosomes. Such trees, it was discovered, grew faster and contained several other characteristics that made them desirable as pulpwood. By the summer of 1955, the Institute had more than five thousand seedlings and grafts with more than the normal number of chromosomes, received from Mexico, Holland, Czechoslovakia, Italy, Korea, Canada, and Sweden. With these experimental plants, Joranson set out to determine if chromosome number could be altered by various influences such as chemicals, heat, and various types of irradiation. As he told his audience at the 1955 Executives; Conference: "We wonder…if it is within the power of man to guide the evolution of certain trees toward properties and characteristics which will be even more useful that those nature has already provided."

The research, academic, and administrative groups at the Institute also underwent substantial changes during the 1950s. At the 1950 Executives' Conference, the Institute's Board of Trustees was completely realigned to bring it to the level of other large educational institutions. Westbrook Steele, previously the Institute's Executive Director, became President, John Strange, previously Secretary/Treasurer, was elected Vice-President, and C.B.

Mead graduated from the Institute in 1952 with his Masters' degree. He immediately went to work at Consolidated as a chemical engineer. "Had I come right out of college, I don't think I'd have felt very prepared," he said. "But having those two years at the Institute gave me a tremendous leg up in terms of credentials and some knowledge about the industry I wouldn't have had otherwise."

Mead continues his relationship with the Institute even today. "I've enjoyed my relationship with the Institute tremendously. I'm proud of it and I hope to see it continue to thrive in its own way. It's been a huge contributor to the industry and a huge contributor to my own success and, as such, I feel obligated to continue my support."

Mead then went on to express his feelings about the Institute in relation to the industry that supports it: "Our industry was pretty technologically inept when the Institute was first formed. While I don't think the Institute or its graduates can take the credit for having created the technological upsurge that has taken place since 1928 or so, the fact is that it couldn't have taken place without the Institute graduates out there to implement it, to help put it into place. The Institute's influence on the industry is probably as great as any institution could have possibly had. Our industry is struggling and it had been for many years. It is comparable to many other basic industries in that respect.

The industry would have been a heck of a lot worse off if it hadn't been for the Institute. I think that's clear. The other institutions overseas and those schools that had at least some paper influence wouldn't have been able to fill that gap. To me, the Institute created the atmosphere of technological aggressiveness which it's been under for a long time now. We haven't solved all the problems by a long shot. We still have a very inefficient papermaking process. There's way too much water, way too much energy and our pulp mills still stink. The process in general is ten times more efficient than it was when the Institute first formed, but it's still got a long way to go. We've come a long way."

Sitterson, previously Assistant Secretary, was elected secretary. Ernst Mahler, who had been the President of the Board of Trustees since the Institute's inception, was given the title of Chairman of the board, while D.C. Everest and John L. Riegel became the Vice-Chairmen.

The 1950s saw the formation of several new academic groups that served to enhance the education of the future leaders of the pulp and paper industry. In 1951, the Carbohydrate Group was formed under the direction of Dr. E.V. "Jack" White, an expert in the hemicellulose and plant gum field. In 1953, the Graphic Arts Section was created with Dr. George Sears as its chief. Sears' repositioning led to the appointment of John Swanson as the leader of the Physical Chemistry Group, which was renamed the Colloidal Chemistry Group shortly thereafter. Finally, in 1955, Dr. A. Neil McLeod, who received his PhD from Cornell University, joined the staff as an economist, a position designed to help the students and staff on matters relating to the fields of economics and marketing.

These new staff appointments and new academic groups led to yet another revision of the Institute's curriculum. First-year students continued their strong emphasis on the theoretical foundation of the sciences. Second-year students continued these basic studies, but added the processes of the pulp and paper industry to their coursework. In 1955, a new course titled "Wood as a Raw Material," focusing on the biological, physical, and chemical properties of wood, was added to the second-year curriculum, along with a new requirement that second-year students take on a field of special study as a preliminary to the doctoral program. The third-year program consisted of three sets of technical problems of a broad and diversified nature designed to give students an opportunity to demonstrate their ability to analyze situations known to exist in industry. The fourth year program was devoted to the student's original research.

Mr. W. Irving Osborne, Jr.

Dr. David L. Luke

Dr. A. Neil McLeod

Dr. Ernst Mahler

Dr. Harry F. Lewis

Dr. Douglas M. Knight

Dr. Roy Whitney

Mr. John Strange

If Otto Kress's 1948 retirement constituted the closing of the first chapter of the Institute's history, then the period from 1954 to 1957 might well be considered the closing of the second. In February of 1954, Dr. Douglas Maitland Knight, formerly a professor of English Literature at Yale University, was installed as the eleventh president of Lawrence College, succeeding Dr. Nathan Pusey, who was chosen to serve as president of Harvard.

In 1955, Westbrook Steele announced his retirement as the Institute's President. Chosen as his successor was John G. Strange, who had served the Institute in a variety of capacities since the early 1930s. Six months later, in March of 1956, Dr. Harry Lewis announced his retirement as the Institute's Dean. Dr. Roy Whitney was chosen to succeed Lewis.

Concluding the wave of retirements was that of Ernst Mahler as the Chairman of the Institute's Board of Trustees in 1957. He was succeeded by W. Irving Osborne Jr., President of Cornell Paperboard Products Company in Milwaukee. David L. Luke, President of the West Virginia Pulp and Paper Company, was named Vice-Chairman.

The Institute truly experienced a "Coming of Age" – as the theme of the 1950 Executives' Conference predicted – in the 1950s. By the end of the decade, the Institute was supported by one hundred and twenty-eight member companies paying more than $720,000 a year in dues. Two hundred and sixty-three staff members worked in offices and laboratories valued at more than $3

An Institute education doesn't guarantee anything – nothing ever does - but it does open doors.

— **Thomas Czepiel,** entering class 1954

I think it gave me a good foundation for going out and accomplishing something and contributing something to the companies I worked for. I appreciated that. I had knowledge of the industry and the chemistry that was involved in the process. I wasn't too self-assured when I finished college, but after going to the Institute, I felt I knew what I was doing. All the relationships that were built there were lasting.

— **Austin Anthis,** entering class 1950

Mr. Westbrook Steele

Dr. Nathan M. Pusey

million. The Institute's library held more than twenty-one thousand bound volumes and subscribed to four hundred periodicals in seventeen languages.

The 1950s also saw the Institute's alumni beginning to seek greater involvement. In 1953, a new recruiting program allowed alumni to speak at colleges and universities on the Institute's behalf. In 1953, the Institute's administrative offices were delighted to hear of the organization of an alumni association by the alumni themselves. The alumni association, still active today, served as an important vehicle of communication between alumni and the Institute as well as providing a mechanism for the nomination of alumni trustees.

Yet through it all, the primary focus of the Institute remained true to its vision: attracting promising students, guiding them, educating them, and placing them in the industry. "If we succeed in that," Mahler promised, "we shall succeed in all."

Irwin A. Pearl

A native of Seattle, Irwin Pearl received his bachelor's, master's, and doctoral degrees in chemistry from the University of Washington. He completed some post-doctoral work in lignin with Dr. H. K. Benson at the university before coming to the Institute in 1941 as a research associate.

As a researcher at the Institute, Pearl worked extensively in the fields of stream pollution and by-product utilization. In 1955, he succeeded Dr. Fritz Brauns as the leader of the Lignin Group. He retired from the Institute in 1976 as a senior research associate in the division of natural materials and systems and a professor of chemistry.

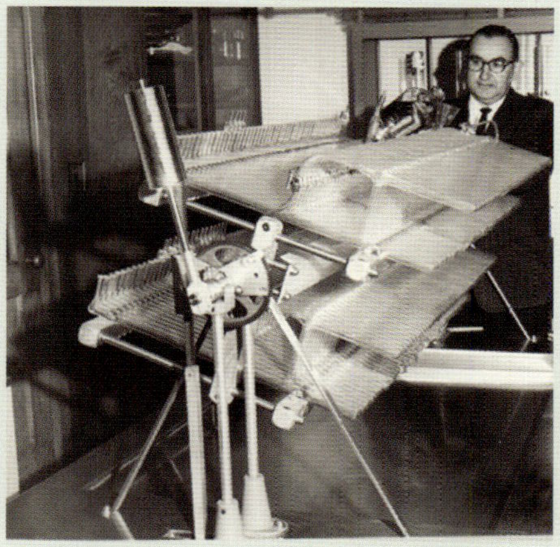

George G. Jernegan

George Jernegan received a Bachelor of Arts degree from Beloit College in 1943 and a Master of Arts degree from the University of Iowa in 1947. He joined the Institute staff in September of 1954 as a staff associate in the Dean's office, coming from Beloit College, where he was an assistant professor of Education and Psychology. Jernegan familiarized himself with the paper industry early in his career and was a major contributor to a publication entitled *A Guide to Career Opportunities in the Paper Industry.*

As a member of the Dean's office, Jernegan was responsible for a variety of extracurricular academic programs such as industry seminars and the summer Fiber Microscopy course. He advanced to the position of director of Continuing Education in 1958. In 1960, Jernegan succeeded Edwin Schoenberger as the Dean of Students, a position he held until his death in 1963. Jernegan was awarded Emeritus status posthumously.

Said John Strange during a memorial service held for Jernegan: "He will long remain among us through memories of his companionship, through his contributions to our program, and through the continuing work of students whose lives he touched so frequently and so positively."

Those were great days and those were great people. The teachers were a great influence on all of us. I've enjoyed my relationship with the Institute tremendously. I'm proud of it and I hope to see it continue to thrive in its own way. It's been a huge contributor to the industry and a huge contributor to my own success.

— **George Mead**, entering class 1950

Johannes Van den Akker

Johannes Archibald Van den Akker was born in Los Angeles, California, on December 5, 1904. He received his bachelor of science degree in physics with an engineering minor from the California Institute of Technology in Pasadena in 1926; he was awarded his PhD from that same institution in 1931 for his work in determining the correctness of quantum theory in the domain of heavy metals and energetic photoelectrons. He continued this research first as a postdoctoral student at Caltech, then as a researcher and instructor in physics at Washington University in St. Louis.

In 1935, "Van," as he was affectionately called by his students and colleagues, joined the staff of the Institute as the chair of the Physics Section, a position he held until his retirement in 1970. In addition to those responsibilities, Van also served as chair of the Mathematics Section during his last fourteen years at IPC.

Van's research interests were wide and varied. Considered one of the country's – and indeed, the world's – top physicists, his areas of expertise were in the fields of optics and thermodynamics. His research in super calendering, mechanical properties of individual pulp fibers, mechanical properties of paper webs, and the optical properties of paper proved vitally important to the United States paper industry.

Van officially retired from the Institute in 1970 and was honored with the title of Professor Emeritus. From 1971 to 1982, he was a research and development consultant for the American Can Company in Neenah; he also served as a consultant for the James River Corporation from 1982 to 1985.

Van's career was filled with numerous honors. He spent 1961 and 1962 in England as a Fulbright Scholar at the University of Manchester Institute of Science and Technology. TAPPI, an organization to which Van was extremely devoted, awarded the physicist the 1967 Research and Development Award as well as the 1968 Gold Medal. He co-authored ten books, contributed to the first and second editions of the *Encyclopedia of Physics*, and penned more than one hundred articles for various scientific and technical journals.

Van den Akker…he was certainly one of my favorites. He was brilliant…very congenial and an excellent instructor. His courses were difficult. You had to study very hard, but he made them interesting. He was one of the favorites.

— Joe Atchinson, entering class 1938

Van had a lot of fun in him – he was a good party man. He was always invited to the student parties.I also played a lot of four-wall handball with Van. I had the great embarrassment one day to hit him with the ball. Hard driven, in the back of the neck – I knocked him out cold. I picked him up and he woke up and said something like, "don't worry – I'm not going to flunk you for this."

— George Ingle, entering class 1938

Van den Akker was one of the great physicists of the world in my mind.

— Daniel Adams, entering class 1939

He was a great one. He was a great personal friend to me later on. I had left a fraternity pin on a cute blonde at Rice and, after seven months, I got a Dear John letter and I was so broken up that I lost interest in the work. Van den Akker took me aside and talked to me like a Dutch uncle and said we all had to hit some bumps in life and we had to pick up and keep going. He was a friend through life and I remember sending him a card on his ninetieth birthday. He lived to a ripe old age. He was outstanding.

— George Martin, entering class 1941

Dr. Van Dan Akker was the most rigorous professor… he gave very good courses but they were tough.

— Richard Brown, entering class 1946

Van was just a tremendously intelligent and demanding guy. He was the filter. As you got halfway through his course he threw a tough test at you and you found out who were the sheep and who were the goats right off the bat. As far as I could tell, that was the one key element that sorted the class out as to who was going to stay and who wasn't.

— George Mead, entering class 1950

I remember Van's TV antenna. In those days, there were no television stations up in the Fox River valley area. The nearest stations were in Milwaukee, about 100 miles away. So you had to put up a high tower to put up an antenna to receive the TV, which probably only came in about 3 hours a night anyway. So Van designed one himself – a steel structure – he basically enlisted a few of us to help him stand it up erect and get it set up. So it worked for six months or a year, perhaps, and then we had a really heavy windstorm come through and the thing blew down. When we talked about how it was too bad the tower blew down, he said, "You know, Art, you can't trust the weather bureau." I said, "What do you mean, Van? A big storm came up." He said, "No! I designed that tower to withstand seventy-five mile an hour winds. They said the maximum gusts were sixty-seven, but I know they were wrong, because the tower would still be up if we hadn't had gusts over sixty-seven miles an hour." I thought to myself, wow, he's confident in his computations.

— Arthur Dreshfield, entering class 1951

I thought he was a wonderful person and a wonderful teacher. As a thesis advisor, he was first-class because he left me pretty much alone, and I appreciated that. If I needed help, though, he was always available.

— John Schulz, entering class 1955

There was something about the way he challenged you with problems, problems that made you want to go the extra mile and try to understand and delve into the problems he gave to you. I found him particularly valuable and enjoyable.

— David Fary, entering class 1960

I learned so much from him. I learned how to write from him and I learned how to research from him. He is such a classical individual. He was a very tough teacher, almost severe. I heard before I came to the Institute that if you can make it through Van den Akker's class, then you can make it through the Institute. I thought he was very delightful.

— Harry Spiegelberg, entering class 1961

Van den Akker was very interesting and dynamic person. He always made tests that made you think a lot because he didn't ask questions that he talked about in classes.

He always came up with different things that made you test the limits of what you had been taught and how to put things together.

— Arne Ahlen, entering class 1963

He was the best teacher I ever had – ever, anywhere. He was able to describe relatively complex things, and sometimes very complex things, so clearly. When explaining a theory, he was always able to bring it down to practice. He was so open, although in a way he was kind of a formal guy...he always made you feel that he was open to you at any time ...he was so knowledgeable and a world-wide expert....one of the guys that seemed to literally have it all, seemed to have no weakness.

— Jim Farrand, entering class 1964

My favorite was Van den Akker - he was absolutely terrific. He was a fundamentalist all the way – "think through your problems from the base up and work it out."

— David Dumbleton, entering class 1966

You would walk out of Van's class knowing that you really learned something that was very important and you could actually use what you'd learned.

— Jim Ferris, entering class 1967

Van den Akker - he was a super gentleman, a very knowledgeable person and a very good teacher. I really owe a lot to him.

— A. Nurhan Becidyar, entering class 1969

Kyle Ward, Jr.

Originally from Beaumont, Texas, Kyle Ward, Jr., received his Bachelor of Science degree from the University of Texas in 1923, his Master of Science degree from George Washington University in 1926, and his PhD from the Friedrich Wilhelm University in Berlin in 1932, where he studied under Professor Pringsheim, one of the world's eminent sugar chemists. Upon his graduation, Ward worked as a research chemist for Hercules Power Company in Wilmington, Delaware, for several years before joining the staff of the Southern Regional Research Laboratory in New Orleans as the head of the Cotton Fibers Division in 1939. He held this position until he came to the Institute in 1951 as Dr. Emil Heuser's successor in the Cellulose Group.

Most of Ward's research was conducted in the fields of cellulose and textiles, especially cotton. He is credited with synthesizing the first member of the nitrogen mustard group, chemical agents now used to treat certain types of cancer. He also conducted pioneer research on naval stores such as turpentines, gums and resins.

Ward's career was filled with awards and honors. In 1945, he received a certificate of appreciation for services with the Technical Industrial Intelligence Committee of the Joint Chiefs of Staff during World War II, and he was named one of the top ten cellulose chemists in the United States by the Chicago Section of the American Chemical Society in 1947. He was named a TAPPI Fellow in December of 1973 and was awarded the Anselme Payen Award of the Cellulose Paper and Textile Division of the American Chemical Society in 1977. Ward also authored numerous scientific papers and published and contributed to several books in area of cellulose chemistry.

In 1966, Ward succeed Dr. Bertie Browning as the chairman of the Organic Chemistry Section. He held this position until late 1967, when he turned it over to Dr. John Green. Ward continued to serve the Institute in a variety of staff interests and activities until his official retirement in October of 1968. Ward was honored with Emeritus status, which he maintained until his death in 1997.

Kyle Ward - he was a cellulose chemist, an author and just about everything else. He was a wonderful man.

— Richard Barker, entering class 1958

The IPC:

An Epic in the Honor of Its 25th Birthday

In '29, housed in the Gym,
Our space and manpower were quite slim.
But now in 1954
We house 200 souls or more.
And tho' we've really grown, I'd say
Please let us not forget the day
When we were insecure and small
Or brag too much because we're tall.
Since ones from tiny seedlings grew
It's well to prize those seedlings too.

And now I trust my withered Muse
Won't bore you, but will you amuse
With lyre, cymbal, drum, and flute
She'll first salute the Institute
By calling for three rousing cheers
For our few early pioneers.

Her first verse might be aimed I guess
At that stout warrior OTTO KRESS
Whose forthright personality
Oft served to boost the I.P.C.
He'd undertake most any chore,
And then find time for many more,
(Provided he'd be found at home
Because quite often Doc would roam.
And even now, his itching foot
In Florida can ne'er stay put.)

When restive Kress moved o'er the map
Dean HARRY LEWIS took the rap
As mate, out in our storm-tossed boat,
Helping to keep the "kraft" afloat
So – a salaam to H.F.L.
Who helped to man that bark *so well.*

An accolade to WESTBROOK STEELE
Whose energy and tireless zeal
Kept everything on even keel;
Whose words, we know, had wide appeal
To those who in the industry
Sponsored out youthful I.P.C.

Our prexy flew o'er dale and hill
To gain another member mill.
His batting average proved sky high
For seldom did a mill slip by.

More often staying within range
Was our kind Veep – I mean JOHN STRANGE
Who serves as treasurer as well,
Which from our point-of-view is swell
Because with Strange up near our head
We'll never fear the taint of Red.

Our secretary is C.B.
Tho' ne'er a Navy man was he.
Perhaps you'd think he'd be *at sea*
When projects flood the I.P.C.
Yes - problems come in by the score
And, when they're solved he garners more.
(With all those items on the beach,
One wonders who finds time to teach.)

Yet – graduate work here is unique
With three fine DEANS up at the peak.
The teacher's rank is somewhat lesser,
For you'll find nary a professor.

The title which the teachers rate
Is termed RESEARCH ASSOCIATE.
(And yet kind friends – what's in a name?
We seem to get there just the same.)

Who worries about heat and light,
And whether things are going right?
Who guards the portals of the till?
Who keeps fool-items off the bill?
Quietly sitting in his den
THOMAS B. WADSWORTH –
known as Ben.
Problems increasing with our pace
He meets with dignity and grace,
A nice bouquet should go to thee –
Hardworking, gentle Thomas B.

A mighty floral horseshoe should
Be handed to the pulping brood
MACLAURIN – JOHNNY PECKHAM –
STONE –
HELLEUR *– e'en tho' we often moan*
When eating lunch – and get their smell
Arising from that lower well.

We'd like to mention everyone,
But that would hardly prove much fun.
And it would be a scurvy trick
To let this issue get too thick.

Our students all deserve great praise
For working nights as well as days,
Making their work a source of pride
While raising families on the side.

Alumni, too, in high repute
Strengthen the Paper Institute.

So orchids to our colleagues all
Whether they're short or rather tall.
Whether they're stout or somewhat lean,
Whether they're slow or rather keen,
Whether they're young or old,
Whether they're modest or too bold,
Whether their hair be black or red,
Or if but a few hairs grace their head.

The chemists and the analysts,
The physicists, biologists,
Cambium layers, geneticists,
Graphic artists, economists.
(Now isn't that a motley crew?
I'm sure my Muse has skipped a few.)

The maintenance crews, the electricians,
The painters, amateur musicians,
The messengers, the stock-room boys,
The janitors – those men of Roy's,
Howell's plasticians, Bob's containers
(And all their various retainers.)

The editors, photographer
And every bibliographer,
The secretaries (Delia's ladies
Who would consign our scrawls to Hades)
Fliers and Coders daren't be missed,
Nor our patient receptionist.
Our business group, librarians –
(Yes, even those barbarians
Who smoke under NO SMOKING *signs.*
Orchids to them! I mean big fines.*)*

My Muse would mention those few bores
Who throw waste paper on the floors,
Or leave the lights on (which I turn off)
(But now the party's getting rough.)
Even those sinners are nice folks
Who litter rooms while drinking cokes.

These carping lines may leave a pall
We'd best hand orchids to THEM ALL!
Then let's join hands in unity
And further this the I.P.C.

- By "Sapiens"
May 1954
The Insti-Tooter

We were blessed with people who were very good at what they do. They were really inspirational to the students.

— **William Trice,**
entering class 1958

Outside the Classroom

In the spring and the fall, I played outside - baseball, football, things like that. I loved to sail. A guy named Jerry had a sailboat on Lake Winnebago. We often sailed in Neenah Harbor. We gave the boat the nickname "dry lab." When someone asked me where I was going in the afternoon, I would just say that I was going to the lab!

— **Arthur Dreshfield,** entering class 1951

We played ball - I was the shortstop on the softball team. A friend and I bought a raft for $40 and we had many memorable trips on the river in it. We did that whenever we could get away.

— **Richard Barker,** entering class 1958

We had a pretty heavy work schedule, but there was always time for something, particularly when I stayed in Kress House. I'd say most every weekend, except maybe when exams were pretty heavy, we'd have a party of some sort.

— **John Schulz,** entering class 1955

I was married, so my social life was based around the married students. We went on fishing trips and I learned how to play golf. There was a golf course right behind our apartment. We also played lots of bridge. We were a very close-knit group of students in the married housing. We didn't have money to do much in town. Most of our activities were limited to the campus.

— **Robert Guide,** entering class 1954

We didn't do a lot of stopping in the bars, though I do remember one memorable occasion. I guess it must have been the end of my first year that Leo McDonnell and I decided to go to every bar on College Avenue - up one side and down the other - and have a beer in each one of them.

— **Vincent Russo,** entering class 1952

The student body would organize dances and parties - the single guys in Kress House would have parties and we would go over there. We were much more active in that area than former student bodies had been. A couple of the guys took some of the apples from the old apple orchard and ran them through the refiner and made hard cider.

— **William Trice,** entering class 1958

There were four of us from the South that had an interesting time with the weather up there. We learned to ice skate a bit - threw snowballs and made snowmen. In the wintertime, Lawrence College had an enclosed swimming pool, so we could go swimming. One time I didn't dry my hair and it was sub-zero weather and the water on my hair just froze. That was an interesting experience.

— **Austin Anthis,** entering class 1950

We had an active social life without any money. We went fishing to put meat on the table. Several of us became avid fisherman and we often borrowed the biology department's net to smelt. We all ended up with enough smelt sometimes to last the winter season. Everyone that wanted to had a garden patch and we raised our own vegetables. The most interesting thing was that we made about two hundred gallons of hard cider using the pulp laboratory. Student housing was in the midst of apple orchard and the apples weren't good to eat. One year we gathered about two hundred pounds of apples, then went to the lab and used refining gear to get many gallons of juice. We set up a fermentation house in the basement of Kress House and produced enough to fuel a New Year's Eve party.

— **John Vaughan,** entering class 1958

We did play cards sometimes and the married couples would get together. We did a lot of fishing after we finished working on our theses. I would go to the lake with my friend and study and fish at the same time.

— **Joseph Koleske,** entering class 1958

I remember making apple cider – hard apple cider – out of apples from the apple trees over at West Court and using laboratory equipment to make it. We would throw a big party for ourselves and the faculty over at Kress House with the things we produced. We had a pretty good time. I remember having a pretty good New Year's Eve party the second year we were there. I was married by that time. We had it down at Astor Hill House. We decided we were going to burn the Christmas tree. Well, the thing got away from us and fortunately our wives were smarter than we were and threw baking soda on it. We would have burnt the house down, trying to shove it into the fireplace.

— **Robert Jones,** entering class 1957

The Bush Trip in the 1950s

I remember a number of mills we visited. They would usually treat us very well because they were trying to leave a good impression with some of the students because some of them might want to work there later on.

— **Austin Anthis,** entering class 1950

We got our families settled in Appleton and then we headed up north. We stayed at the Trees for Tomorrow camp in Eagle River. Isenberg taught a course in wood identification. We also got acclimated to forestry practices in a very limited but hands-on way. We visited a whole bunch of various wood consumers, a lumber mill, and several pulp and paper mills throughout the area. That whole trip was very worthwhile. Besides that, it gave us a chance to get together, the fifteen of us, and get to know each other and that was a really important factor too.

— **George Mead,** entering class 1950

The bush trip was certainly a neat introduction. It was so different from anything we'd had in undergraduate school. The subject matter was interesting, but going up to a beautiful part of northern Wisconsin was just great. Being allowed to see first hand what a forest was like, not just studying it academically – how the hardwoods and the softwoods all fit together – was a plus.

— **Arthur Dreshfield,** entering class 1951

One of the class mottos was developed in the Trees for Tomorrow. When in doubt, sack out - forget it and take a nap. There are certain things you can do and certain things you can't do - sometimes you quit trying and take a nap. When you wake up, you take a new perspective.

— **John Parkinson,** entering class 1951

It was a fine education. I was really happy with the career I had as a result - the work was interesting and challenging.

— **Arthur Dreshfield,** entering class 1951

It was a unique type of education that prepared me to move ahead in the industry. It taught us how to think for ourselves, how to attack problems and how to go after them and deliver them. We weren't going out to be scientists; we were going out to be technical managers. It was a tough grind, but I think they were some of the best years of my life. I have very positive memories from it even though it was very hard work.

— **Robert Jones,** entering class 1957

We got to visit a number of mills and get a sense of what the industry was like. It was a camp-like setting, so we played a little softball in the afternoons, which was fun. I remember my glasses getting broken. I broke my glasses the first day out and we had to find a place in Eagle River that could piece them together.

— **Vincent Russo,** entering class 1952

I went there knowing nothing about the Institute or about the paper industry. Many of the other guys there were very familiar with it. Some of them had worked in paper mills or had come from paper families. I had a lot of adjusting to do. It was a good learning experience and a chance to get to know the classmates.

— **John Schulz,** entering class 1955

The togetherness - we got know each other really well.

— **Richard Barker,** entering class 1958

We bonded pretty well. After touring the forestry land and the water resources, we went fishing. When the sun went down, we got the bus driver to take us to some bars around town. He would pick us up at midnight and we would start all over again the next day. One night we fried all the fish we saved and had a beer bash.

— **William Trice,** entering class 1958

I thought it was one of the best things we could have done. There were twenty-five of us in the class and it was nice to start classes knowing one another. About half of the class had been out from undergrad for a few years. About half were married. The week helped bond us into a cohesive group.

— **John Vaughan,** entering class *1958*

CHAPTER 4

1960-1969

We stand today on the edge of a new frontier . . . a frontier of unknown opportunities and perils, a frontier of unfulfilled hopes and threats. The new frontier of which I speak is not a set of promises—it is a set of challenges.

— **John F. Kennedy,** Los Angeles, California, July 15, 1960

Some call it the best of times, others the worst of times, but one thing is certain – the 1960s changed America forever. Radical activism swept the country, bringing with it a restructuring of American society, politics, and culture. The 1960s also saw a revitalization of political and cultural conservatism that paved the way for our modern political arrangement. Above all, it was the age of youth, with seventy million children of the post-World War II baby boom com ing of age and desiring change – whatever the direction.

The Institute of Paper Chemistry – at times, seemingly a world unto itself – was also standing, as President Kennedy so eloquently stated, "on the edge of a new frontier." No longer a child in the worrisome Depression years, no longer an awkward teenager during the wartime years, no longer a cocky young adult during the 1950s – the Institute was now fully grown and entering its heyday. Despite the string of retirements that characterized the end of the previous decade, the In stitute's staff still boasted some of the most prominent scientific minds in the country. Those minds were presented the challenge of not only educating the "best and the brightest" who came to Appleton to study pulp and paper technology but also to conduct research that would lead the American pulp and paper industry to the edge of a new frontier as well…

The Institute of Paper Chemistry entered the 1960s with the same attitudes of hope and optimism that permeated the rest of the country. A fresh, young president dared the nation to face the challenges posed by the arrival of the new decade with courage and dignity – to "ask not what your country can do for you - ask what you can do for your country." In Appleton, Institute President John Strange issued a similar challenge to faculty and staff during his speech to the 1960 Executives' Conference: "to continue and strengthen our research efforts…to equip ourselves with the latest in scientific tools…to reach a state of intellectual readiness for the more difficult questions and challenges which are not so very far down the road."

THE BUILDING BOOM of the 1950s did not abate with the dawning of the new decade. In late 1961, plans were announced for a new education and activities center to be built on West Avenue. The new building was made possible by a $325,000 grant from the Louis Calder Foundation, which had also donated funds for the Louis Calder Plant Biochemistry Lab, an electron microscope and auxiliary equipment, and the Lou Calder Scholarship Fund.

Construction on the Lou Calder Student Center began in March of 1962. The building was comprised of two wings. The first was an assembly hall capable of seating two hundred people for seminars and conferences or one hundred for dining. The second wing consisted of a two-story housing unit containing twenty-four single rooms, a small kitchen, and a dining room. The housing wing provided residence for the Institute's single students as well as guest rooms for visiting scientists and lecturers. A communal lounge and lobby

John W. Green

Dr. John Green received his undergraduate degree from Princeton University and his Master of Science degree from the University of Pennsylvania. He then returned to Princeton to complete his PhD in organic chemistry. He joined the staff of the Institute of Paper Chemistry in 1938 as a member of Dr. Emil Heuser's Cellulose Group.

During World War II, Green, along with the majority of his IPC colleagues, focused his attention on research projects directly relating to wartime activities. After the war, he began to explore what would become his lifelong research interest – simple sugars. Throughout the 1950s and 1960s Green's research centered on the not-so-simple acids formed from sugars in alkaline pulp liquors.

Green was named a Senior Research Associate in 1966. In 1967, he was chosen to succeed Dr. Kyle Ward as the Chairman of the Institute's Organic Chemistry Section, a position he maintained until his retirement in 1977.

John Green was a great guy. I got to know him fairly well when I was taking his analytical chemistry class. There's a story kicking around out there about me trying to dry out a solvent solution in one of John's ovens and blowing the oven up, which is true. I did it.

— George Mead, entering class 1950

John Green was always interested in students' ideas and was always happy to sit down and talk about something - not only technical things but things about life in general. He was always available.

— John Schulz, entering class 1955

He was a very personable guy. He got along well with the students. We were chemical engineers and he was teaching us organic - he was very helpful in working us through that. There was a great rapport between John Green and his students.

— Robert Jones, entering class 1957

area joined the two wings. The building, completed in the fall of 1962, was designed to be a harmonious part of West Court Village, which was once again enlarged by the construction of the tenth student apartment building during that same year.

In May of 1967, the Institute's trustees approved a $5,000,000 development program designed to increase graduate student enrollment by one-third, improve the quality of the graduate education program, improve, extend, and accelerate research activity, improve and greatly expand activities relating to the collection, storage, retrieval and dissemination of scientific and technological information, place the continuing education program on a year-round schedule, and, ultimately, to carry forward more vigorously the long-range programs of education and research which the Institute had undertaken on behalf of a group of growing industries.

The program commenced in the summer of 1967 when the Inland Container Corporation made a gift of $1,000,000, the largest single gift in the Institute's history. The money was to be applied toward the cost of building and equipping a $1.5 million graduate research center to be used for research and research training of students admitted to doctoral candidacy. The thirty-two thousand square-foot building, named the Krannert Graduate Research Center in honor of Herman C. Krannert, who founded Inland Container in 1925, also featured an auditorium with a seating capacity of three hundred and sixty. The building was dedicated during the 1969 Executives' Conference.

The new auditorium allowed the Institute to move forward in another phase of the development program, which involved transforming the old auditorium in the General Activities Building into a computer center. The center, featuring a "third generation" IBM System 360-Model 44 and complementary equipment, was completed in December of 1968 at a cost of $600,000. The new computer system was a vast

That was a really good time in my life.

— **Harry Spiegelberg,** entering class 1961

Herman C. Krannert

improvement over the IBM 1620 that had previously been used by the Institute, and the new computer center greatly expanded and enhanced the processing, retrieval, and utilization of scientific information.

The final stage of the development program was the construction of a new nineteen thousand square-foot continuing education center, which began in July of 1969. Designed to put the continuing education program on a year-round schedule, the new building was outfitted with a large seminar room, two small seminar rooms, a lounge area, a kitchen, dining facilities, and twenty-eight dormitory rooms.

For me, it was a big benefit because it forced me to think in terms of systems that were very broad. Also, getting through it gave you a sense of confidence that there wasn't any technical area that you couldn't deal with. It kept me from being intimidated by new subjects.

— David Eary, entering class 1960

The array of construction projects throughout the 1960s were necessary to create the space needed for the numerous research projects undertaken during the decade. On April 1, 1960, a $600,000 pioneering research program underwritten by the pulp and paper industry and led by the Institute of Paper Chemistry commenced. The program, which quickly gained international recognition, awarded grants to scientific researchers in various parts of the world for long-range and pioneering research studies that held potential for importance to the pulp and paper industry.

The resulting research placed a strong emphasis on high polymers. Studies were grouped into four categories. The first contained three studies on biological, biochemical, and biophysical concepts and theories as applied to natural high polymers that took place at the Institute, at Lawrence College, and at Yale University. The second category consisted of seven studies focused on the organic chemistry of certain natural polymers, particularly the polysaccharides. Two of the studies took place at the Institute; the other five took place at the University of British Columbia, Ohio State University, the University of Edinburgh, the Royal Institute of Technology at Stockholm, and Purdue University.

The third category contained two studies in the physical chemistry of both natural and synthetic polymers, one at the Pulp and Paper Research Institute of Canada and one at Rensselaer Polytechnic Institute. Finally, the fourth category was made up of three studies in basic engineering and technology conducted at New York University, the Massachusetts Institute of Technology, and the Finnish Pulp and Paper Research Institute at Helsinki.

The Institute embarked on many more important research projects throughout the 1960s. In 1960, a joint project with the Nebraska Agricultural Research Fund Committee was announced. A series of three separate projects aimed at

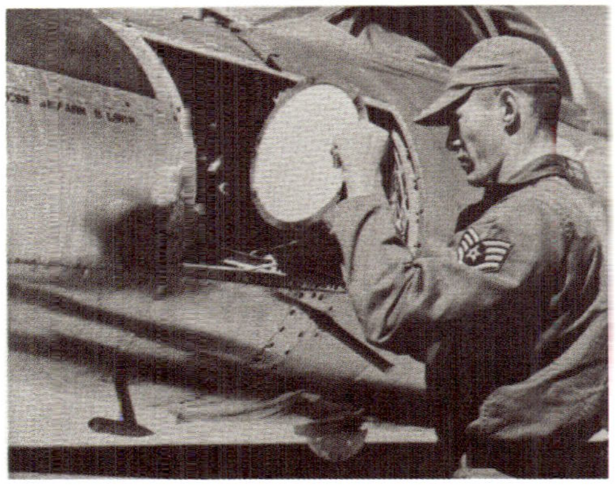

It couldn't have been better - it was the way a university education ought to be. It was a great opportunity to mix with an intellectual community.

— David Dumbleton, entering class 1966

preserving agriculture's present starch market in the paper industry took place over the course of a year under the direction of Dr. John Swanson, the head of the Institute's Physical Chemistry Section.

In May of 1963, a project that had been initiated in 1949 under the direction of Dr. Roy Whitney, Dr. Johannes Van den Akker, and Dr. Leonard Dearth but given classified status by the United State government, gained national recognition. IPC 1478 filter paper - small pieces of felt-like paper that enables scientists to measure and predict "fallout" from nuclear explosions – was researched, developed, and hand-produced by Institute researchers at the request of the federal government as part of a program of sampling the atmosphere for radioactive debris from nuclear explosions. The paper, which could filter radioactive particles about one micron (forty millionths of a inch) from the atmosphere, was tested during "Project Ashcan," a series of test runs performed by the Defense Atomic Support Agency. The tests involved sampling runs U-2 aircraft up to seventy thousand feet, balloon runs up to one hundred and twenty thousand feet, and a space probe run up to two hundred and fifty thousand feet.

Dr. Dean Einspahr

It was a multi-disciplined education. We were focused toward the industry, sure, but we were not focused toward a single discipline. We took physics, chemistry, forestry, mechanics – you name it.

— **Bob Moran**, entering class 1966

Edgar E. Dickey

Originally from Cleveland, Ohio, Ed Dickey received his Bachelor of Science degree from Ohio Wesleyan in 1934 and his Master of Science degree in chemistry from Ohio State University in 1941. He completed his PhD in carbohydrate chemistry while on the Ohio State staff as a member of the university's Research Foundation. Dickey joined the Institute staff in 1949 to work in what was at that time considered the brand-new field of chromatography.

By the mid-1950s, Dickey was considered an expert in radiochemistry – the tagging of various compounds with radioisotopes and following the resultant radio-activity in reactions. Citing his impressive teaching record at the Institute, membership on students' special studies and doctoral thesis committees, and various academic committees as well as his familiarity with the industry, the Institute's Board of Trustees named Dickey Dean of Students in 1963. He maintained this position until 1966, when he was succeeded by John Parker. Dickey returned to his previous position as a Senior Research Associate in the Organic Chemistry Section. He retired in October of 1976 and was honored with Emeritus status.

You say "the isolation's tricky?"
Just hand it to E E. Dickey!
In lecture room or seminar,
Ed's output is far above "par"
His background's wide in chemistry,
In humor, and humanity.
Often on Ed students depend,
Because he is their selfless friend
For colleagues, he sees problems through
Which lesser men could never do
We're lucky to have E.E.D
Upon our happy faculty.
— The Institooter, March 1952

There was a couple - Mr. and Mrs. Dickey. He was the Dean of Students. They were such a warm couple, very inviting and friendly and open

— Richard Erickson, entering class 1955

Ed Dickey was just a wonderful person and really a character. He was certainly one of the high points at the Institute.

— Dick Ottman, entering class 1956

Roy Whitney

Originally from Milo, Maine, Dr. Roy Whitney earned his Bachelor of Science degree from MIT in 1935, then his Master of Science degree from the same institution in 1937. Upon graduation, he joined the staff of the Bangor Station School of Chemical Engineering Practice, where he was first the assistant director, then the director, in charge of a program of plant development work and its relation to the plant operations. It was during this time that Whitney developed a special interest in pulp and paper manufacture.

In 1942, Whitney joined the MIT faculty as an assistant professor of chemical engineering, a position he held until 1945, the same year he completed his doctorate in chemical engineering. After World War II, Whitney became a professor of chemical engineering at the University of Maine at Orono. He was with the university until 1947, when he joined that staff of the Institute as a research associate and head of the chemical engineering program.

In 1956, Whitney succeeded Dr. Harry Lewis as the Dean of the Institute. In 1958, he was elected to the Institute's Board of Trustee as the Vice President of Academic Affairs, a position created in recognition of the growth of the Institute and the importance of its academic program. His progressive educational philosophy - "disciplined intellectual curiosity and the opportunity to exercise it," as he told the 1956 Executives' Conference – was a large contributor to the worldwide recognition the Institute gained throughout the 1950s and 1960s and the high regard with which it was viewed.

Whitney's professional honors were legion. In 1948, he received national recognition for contributions to the profession of chemical engineering by the American Institute of Chemical Engineers. He was named an Outstanding Educator of America for 1974, and he was the recipient of the 1980 TAPPI Gold Medal.

Whitney served as the Institute's Vice President of Academic Affairs until 1977. He then served two years as the assistant to the President before his official retirement in 1979, at which time he was named Professor Emeritus of chemical engineering. During a ceremony honoring his contributions to the Institute's research and academic programs, Whitney spoke to the assembled crowd: "I have thoroughly enjoyed every moment while working here and I will always be interested in the Institute, its people and its progress."

Roy Whitney was a great engineer. He was a super guy and a wonderful teacher.

— George Mead, entering class 1950

Because of his impact on the students, Roy Whitney was as impressive a mentor as anyone ever had. He brought to the students enthusiasm for the industry.

— Vincent Russo, entering class 1952

Roy was an outstanding chemical engineer and an impressive person academically. He was very fair, but demanding. I used the things he taught throughout my life and career.

— Robert Guide, entering class 1954

Dean Whitney was the kind of guy who really looked out for all of us. He was a stern disciplinarian, but he was an excellent teacher, too. He was really an outstanding dean and made the difference for a lot of us in getting through school.

— Robert Jones, entering class 1957

The Forest Tree Genetics program was begun in 1954 under the direction of Dr. Philip Joranson. Financed by Rhinelander Paper Company and Marathon Corporation, the program sought to determine how and to what degree genetic make-up influenced both qualitative and quantitative aspects of the growth of trees. Its goal was to breed a new, improved species of tree that would yield more tons of paper per acre of forest. A series of crossbreeding experiments led to the development of a Bigtooth Hybrid Aspen in the late 1950s.

Dr. Lawson Winton

Dr. George Jernegan

It was a very good maturing process for a guy – an excellent growth experience. It was always a good, comfortable place where you felt secure and you felt part of the group. When you look at the whole society and how impersonal everything's become, you can see that it was one of those last bastions of fellowship, where you developed together and supported each other. The stretch of five years I was there was probably the greatest time of my life... so many neat things happening - my kids were starting to grow up and all the things happening there.

— Jim Farrand, entering class 1964

In May of 1963, the Institute leased from the state of Wisconsin sixty acres of woodlands in the Black River Falls Forest near Millston to be used as a test site for growth of new types of trees developed in Joranson's program, now under the direction of Dr. Dean Einspahr. The test plot was chosen to see how well particular hybrids do in sandy, dry soil; other test sites at Eagle River, Cornucopia, and Marenisco offered other types of soils and environments.

In June of 1963, another test plot was developed in Clintonville. The ten-acre site, selected, like the Millston site, for its sandy soil, was planted with Larch, Bigtooth Aspen, Quaking Aspen, and Bigtooth Aspen crossed with Poplar. Thirty-six trees were planted nine feet apart on the test plots. The Institute set as its goal favorable results within twenty to twenty-five years.

The greatest success of the Institute's Forest Genetics program came in April of 1969 when Dr. Lawson Winton, a member of Einspahr's team, culminated six years of research with the announcement of the world's first "test-tube tree" – the first time a complete tree had been produced by the tissue culture process.

Research on the "super tree" began at the Institute in 1962. Eventually a method to eliminate the need for sexual propagation was discovered, making possible the growing and reproducing cells outside the natural body of an organism.

Institute scientists chose a type of Quaking Aspen tree called the triploid Aspen for their project because of its rapid growth characteristics and its ability to develop new shoots after harvesting. Of the five trees were developed over the course of one year, three survived; two of the three had reached three feet in height by the time the project was announced to the public. It was, according to Arnold Grummer, the Institute's Editor of General Publications and responsible for the Institute's communications with the public, "just wonderful... simply unbelievable."

Despite a long series of successful research efforts throughout the decade, several issues concerned the Institute during the

Dr. Arild Miller

Dr. J. Edward Todd

Dr. Douglas Knight

Dr. Curtis Tarr

Dr. Edgar Dickey

Dr. John Parker

1960s. In May of 1960, the Institute's Board of Trustees learned of a $125,000 deficit from the previous year's $2.1 million operating budget, caused by a growing disparity between membership dues and outlay for educational purposes as well as a transitory lag in cooperative services revenue. Needing a solution to the fiscal crisis, the Board decided to raise membership dues and maximum contracts by twenty-five percent, representing the first general increase in membership dues since Institute was founded.

During that same Executives' Conference, the Institute's Dean, Dr. Roy Whitney, revealed to the Board the difficulties the Institute was having in finding quality students to fill the school's classrooms. The number of students in chemical engineering was declining and the "classical physics" – mechanics, heat, light, and sound – were rapidly disappearing from the academic scene as the "glamour fields" of mathematics and physics were taking over. By 1962, student enrollment had dropped to fifty-four and the Institute was forced to reverse a policy that had been in place since the school's inception – that all students admitted to the Institute were expected to complete the PhD program. The new policy allowed admittance of students who only planned to get a Masters degree.

There were also several staff and administration changes throughout the 1960s. George Jernegan replaced Dr. Edwin Schoenberger as the Dean of Students in 1960, while Dr. Arild J. Miller took over the position of Director of Admissions from Dr. J. Edward Todd, who announced his retirement in 1961. In November of 1962, Dr. Douglas M. Knight, Lawrence's president, was elected president of Duke

I thought the Institute did a good job of preparing people for running an industry that has some growth to it.

— **Jim Burke**, entering class 1961

It was certainly a positive experience. I made a lot of good friends there.

— **Arne Ahlen**, entering class 1963

I am proud to be an Institute graduate. The Institute is famous for producing generalists, not specialists. My life story is about being a generalist and being an inquisitive person. That is why I am in the place that I am in right now. I was taught to think out of the box.

— **A. Nurhan Becidyan**, entering class 1969

University, thus demonstrating once again that Lawrence served as somewhat of a "proving ground" for major university presidents (Dr. Henry Wriston left Lawrence to assume the presidency of Brown University, while Dr. Nathan Pusey left Lawrence to be the president of Harvard). Knight's replacement, Dr. Curtis W. Tarr, previously Assistant Dean of Humanities and Sciences at Stanford University, was chosen in May of 1963.

In the spring of 1963, Dr. Edgar Dickey replaced Jernegan as Dean of Students. He maintained this position until the summer of 1966, when he resigned to focus more on teaching and his research in organic chemistry. He was succeeded by John M. Parker, who served as the Dean of Students until the Institute's move to Atlanta in 1989.

The decade ended with a massive reorganization of the Institute's one hundred and seventy-member staff. Four new research divisions were created, including a division of industrial and environmental systems, a division of materials engineering and processes, a division of natural materials and systems, and a division of informational services. The new system, which allowed research problems to be looked at with a broader view and in relation to other fields of concern in the same division, allowed research to transcend artificial divisions created by the academic disciplines and bring several areas of researchers into the study of one specific field. John Strange explained the new system to the 1969 Executives' Conference: "When a problem cuts across several fields or groups of people - and most of them do nowadays - structuring can be vital to sensible progress."

In addition to the internal issues that the Institute faced throughout the turbulent 1960s, several external factors also gave life in Appleton a little "shake-up." In the wake of the twelve harrowing days of the Cuban Missile Crisis in the fall of 1962, the Institute became one of seventeen private buildings and schools in Outagamie County licensed as public fallout shelters due to its status as a building with a protection factor of one hundred or better, meaning the relation between the amount of radiation from fallout that would be received by a protected person within a shelter compared to an unprotected person outside. The basement areas of the General Activities Building, the Central Research Building, and the South Research Building were stocked with supplies for six hundred and sixty-five people.

The Vietnam War also began to play a role in Institute life in the mid-1960s. By February of 1968, ten of the first-year students – about eighty percent of the class – became eligible for the draft following a new rule that wiped out blanket draft deferments. The deferred status of some three hundred thousand students throughout the country came to an end. Many of these students chose to enlist and get their military service out of the way before starting graduate school. This trend played a significant role in the Institute's recruiting efforts through the duration of the war.

"As most graduate schools turn more and more to the federal government for support and guidance," Dr. Roy Whitney stated, "our devotion to the pulp and paper industry becomes increasingly a mark of distinction and a point of pride... we need students that will bring not only specific knowledge but an ability to look beyond the manifestation to the underlying principles and to utilize these principles in orderly and logical development."

Louis E. Wise

My main interest is in teaching – in helping in the development of young men – not only intellectually, but ethically as well. My relationship with students has usually been quite informal. The men feel free to consult me whenever need arises. They have access to my personal library and to my laboratory and to my time as well.

— Dr. Louis E. Wise, 1947

Louis Elsberg Wise was born in New York City in 1888. After obtaining his Bachelor of Arts degree in 1907, he entered Columbia University, where he earned his PhD in organic chemistry in 1911. He spent the next three years in instructional positions at Columbia and at the University of Missouri before joining the United States Department of Agriculture as a biochemist. After World War I, Wise joined the staff of the New York State College of Forestry at Syracuse as a professor of forest chemistry. He remained at Syracuse until 1932, when ill health forced him to take a leave of absence from teaching. He spent nine years in Florida writing and abstracting.

Wise joined the staff of the Institute in the spring of 1942 as a research associate and instructor in wood chemistry. His research interests were varied; he was particularly interested in cellulose and hemicelluloses with special reference to the isolation and properties of arabogalactan from Larchwood. Wise authored more than one hundred publications in the fields of pure organic chemistry, essential and fatty oils, soil chemistry, biochemistry, wood chemistry, and pulp and paper. With L.F. Hawley of the Forest Products Laboratory, he published one of the definitive volumes in his field, *The Chemistry of Wood*, in 1926.

Wise earned numerous honors throughout the course of his career. In 1944, he was named one of the "American Men in Science" in the field of wood chemistry. He received the TAPPI Gold Medal in 1960, and was the recipient of the first Anselme Payen Award given by the Cellulose, Wood, and Fiber Chemistry Division of the American Chemical Society in 1962. Upon his retirement in 1966, he was awarded a Master of Science *ad eundem* from Lawrence College as well as the honor of being named one of the youngest Emeritus Professors in the history of the Institute. He passed away in 1980 at the age of 92.

Conscience (Study in Grey)

By Louis E. Wise

How lonely are those children's playgrounds;
Their swings deserted;
Slides encrusted with grey snow
Melting to form dark puddles at their base.

These bring to mind our own dark consciences;
Not white like the newly fallen snow,
Not yet graphitic black – but grey.
These serve for most of us.

Few consciences are white,
And, perhaps more are ebony black.
White ones belong to men like Bunche,
Pauling, Luther King, Luthuli.
And a few friends, who all love peace.

These white ones hold no fears,
Nor do they compromise with Truth.
Their owners feel compassion,
And seek to understand what motivates
Those men of bitterness and violence.
Whose dark activities defile the days
And plastic bombs tear through the night.

Deep Purple

By Louis E. Wise

A girl once lay on a purple couch
Her eye held a violet gleam;
And she wrote a wee note on a lavender card
By the light of an amethyst beam.

Can I ever forget that lavender card.
With its weird impassioned caress.
Or the violet gleam, or the amethyst beam,
Or the girl in the Tyrian dress?

Well, all this happened six decades ago,
And of course I can laugh at it now,
But the violet gleam and the lavender card
THEN – brought beads of sweat to my brow.

Louis Wise was an introspective person, very quiet. Very intelligent.

— Bob Boehm, entering class 1948

Dr. Wise could occasionally throw parties for students – always lots of fun.

— John Parkinson, entering class 1951

Louis Wise came out of a tough situation in Europe, so he still did his abstracting by taking old envelopes and opening them and using them as paper to write on. I mean, there we were, surrounded by all kinds of paper and Louis Wise was still being very frugal about the use of paper based on his own experiences.

— Vincent Russo, entering class 1952

I had a fortunate experience in being Dr. Wise's last student. He was the finest gentleman I have ever known as a person, scientist, and humanitarian. He was my mentor. We hit it off personally and professionally. He and his family became friends. Our children saw them as grandparents. We spent time together socially. Dr. Wise was not only a scientist, but also a published poet. He taught me to write. He would read two copies of my quarterly reports - one from a technical standpoint and one from a literary standpoint. That was the better of the two educations I got - writing. He didn't do that for anyone else. It was part of our personal repertoire. It was his gift to me.

— John Vaughan, entering class 1958

Women at the Institute

In September of 1967 the Institute experienced yet another watershed moment: the acceptance of the first female student in the school's thirty-seven year history.

Mrs. Anna Schmierer broke the Institute's longstanding "men only" tradition. Originally from Medford, Massachusetts, Schmierer earned her undergraduate degree in biochemistry from Bates College in 1963. After a stint at Smith Kline, and French Pharmaceuticals, Schmierer – often described as a "pioneering spirit" by friends and colleagues alike – became the first research chemist at the S.D. Warren Company in Westbrook, Maine.

Schmierer's introduction to the Institute came in the form of a casual conversation with John Swanson. She sent in her application and received her acceptance letter just days after her husband, an optometrist, was drafted into the armed services. He went to Vietnam, she came to Appleton.

Schmierer spent one year at the Institute, ultimately deciding to pursue another course of study elsewhere. In an article in the January 21, 1968, issue of the *Appleton Post-Crescent*, Schmierer described her Institute experience: "The students and faculty have always treated me well. They tend to be protective and really concerned that I like it here and do well."

Here, two female students from the 1970s detail their experiences in what had for so long been a man's world.

I didn't realize it at the time, but we were really breaking ground. There were three women in my class – we pretty much doubled the number of women who had ever attended the Institute. We were a watershed year – there were female students every year after that. It was never that big a deal. The guys were used to having women in their undergraduate classes. It was always like one big family – we did everything together.

— **Paula Stoppler,** entering class 1975

I never noticed being treated differently or being made to feel like I shouldn't be there. I was always treated with a great deal of respect. For that time, that was saying something – so many fields were still predominantly male and you had to fight to get anywhere. I never felt that way at the Institute.

— **Gail Carels** entering class 1977

The Bush Trip in the 1960s

I was – and still am - a city boy. I had never spent the night in the forest before. It was primitive, but fun. I had a great time.
— **A. Nurhan Becidyan,** entering class 1969

It was a great opportunity to see what the Wisconsin industry was like.
— **Jim Ferris,** entering class 1967

We went out in the woods a couple times and saw some logging operations. There were trees that the ranger had marked that the logger could take. The logger would go up and fell the tree, then drop the tree so that the butt of the tree was downhill. Then he would go up and knock the big branches off with his axe or chainsaw. After that, he would hitch his horse to the tree and signal the horse. The horse would walk towards the road, dragging the tree down the hill. In the meantime, the logger would go cut down one or two more trees and start knocking the branches off. When he saw the horse was down to the road, he'd go and unhitch it, then bring it back up and hitch it to the next tree. It was so labor intensive – it's absolutely inconceivable today.
— **Al Jones,** entering class 1962

We went up to Eagle River. We played a lot of volleyball when we had free time. Mostly we went around to look at the forestry operations of some of the different companies located there.
— **Jim Burke,** entering class 1961

I enjoyed the opportunity to bond with the other students and learn something in the process.
— **Harry Spiegelberg,** entering class 1961

It was definitely one of the highlights of the Institute. It was an outstanding thing they did because it was the time that really brought us all together. It was a time of real bonding and developing camaraderie.

— **Jim Farrand**, entering class 1964

We went up to Trees for Tomorrow camp with Irv Isenberg. I really enjoyed it. I learned a lot on that trip. It was really good for me to get into the botany of the business and to learn about all those trees and the commercial side of forestry. It was absolutely a great way to begin.

— **David Dumbleton**, entering class 1966

The Institute gave me the combination of a technical and a business background to go directly out of school into an industrial environment and make fairly significant contributions to the industry.

— **Richard Peterson**, entering class 1964

When I left the Institute, graduates were highly regarded and very much in demand throughout the industry. Institute graduates went into pretty responsible positions in a short period of time. It was a wonderful opportunity to get an education and then to be appreciated for what you knew.

I enjoyed my time in Appleton and it certainly led me through a career that I thought was really interesting. I had the chance to do, and learn, and contribute throughout my career. I'm sure I would've done those things in some other industry, but the Institute certainly provided a ready opportunity for that.

— **Al Jones**, entering class 1962

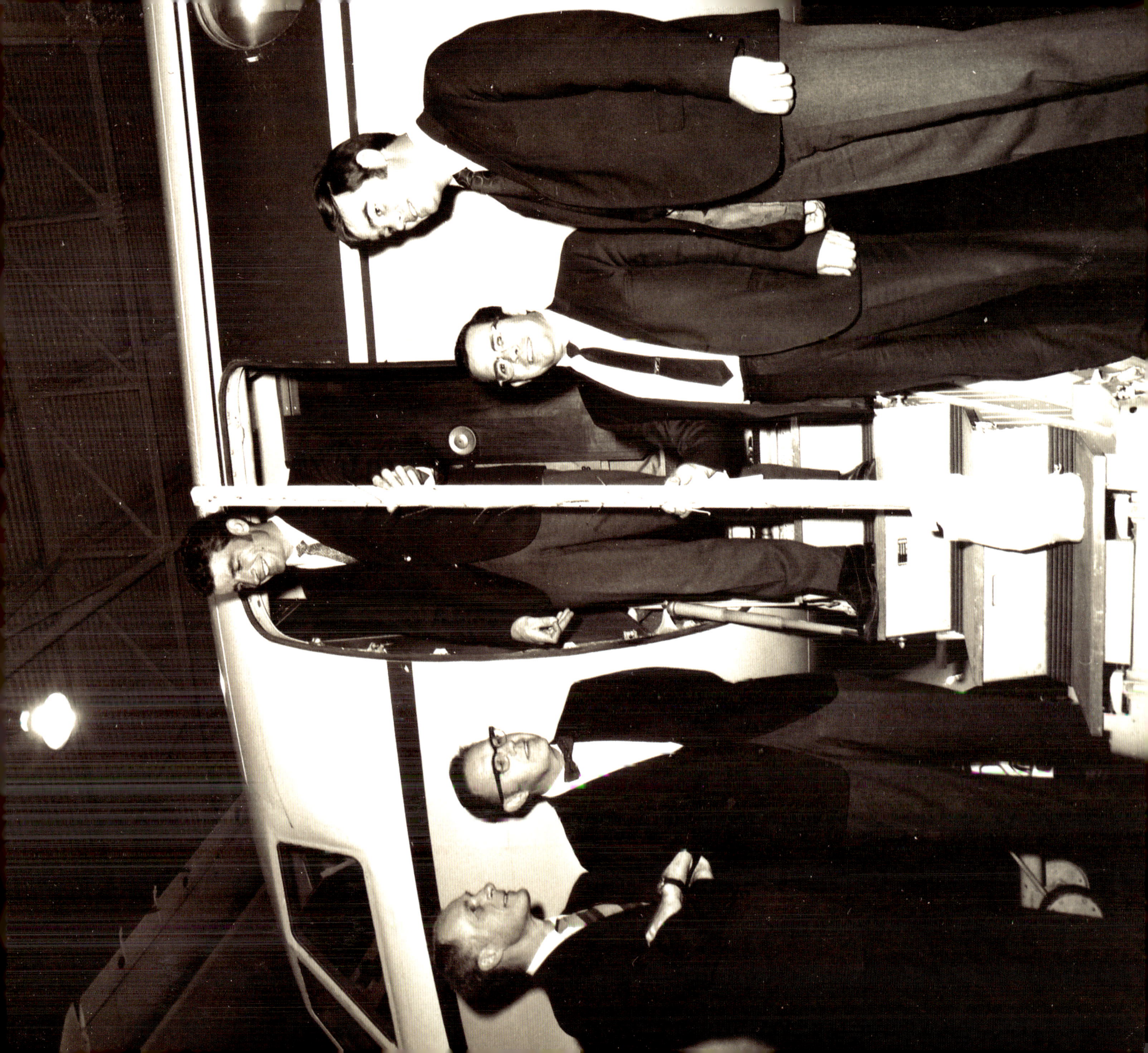

CHAPTER 5

1970-1979

The 1970s saw a continuation of the social activism that so definitely characterized the previous decade.Women's rights, gay rights, environmentalism – everyone had a cause.

Many of these causes played out on the campus of the Institute of Paper Chemistry as well as on the national stage. The 1970s saw the Institute's first female graduate and then the presence of women in each entering class. Environmentalism took center stage for the Institute's scientists, who spent staggering amounts of time on research projects designed to boost environmental awareness in the United States pulp and paper industry…

The 1970s saw a perceptible shift away from the social activism of the 1960s and into social activities designed to heighten one's own pleasure. Indeed, this shift inspired American novelist Tom Wolfe to christen the 1970s "The Me Decade."

One element of social activism that survived this shift, however, was environmentalism. Publications such as Rachel Carson's *Silent Spring* and a series of environmental disasters such as the Cuyahoga River fire and the Love Canal crisis inspired increasing awareness of human threats to the natural world as early as 1962. By the mid-1970s, the majority of Americans were well aware of the damage that industry was doing to air, water, and the earth itself, leading to the creation of governmental agencies such at the Environmental Protection Agency as well as a series of campaigns – such as Earth Day - designed to promote global environmental awareness.

Following the worldwide trend, environmentalism became a key component of the Institute's research program. In 1970, a new division was created for research and development in industrial and environmental systems that proved a significant force in the battle against pollution. A central laboratory for environmental studies was developed to provide an organized systems research approach to industry-wide problems of environmental quality, alternatives in pollution abatement measures, and an organization designed to undertake research for federal and state agencies through contracts, grants, or cost-sharing arrangements.

The Institute's most labor-intensive environmental project of the decade began in 1971 when Institute researchers were asked by five hundred and

The Institute's First Female Graduate

Originally from Maine Marilyn Hobart Kinsey attended the University of Connecticut's School of Engineering for a brief period before settling into the pulp and paper program at the University of Manchester in England. As she told the *Post-Crescent* in a June 1970 interview, "I felt that the bias against women would be less of a factor in the pulp and paper industry." She earned her B.Sc. degree in paper science in 1967.

In 1968, after researching several European pulp and paper graduate programs, including the program at the Norwegian Pulp and Paper Research Institute in Trondheim, Norway, Kinsey made the decision to return to the United States for her graduate education. "I had become so acclimated to living in Europe," she said in a 2006 interview, "that I thought that if I didn't come back to the States at that point in my life, I probably never would." She applied to the Institute of Paper Chemistry and was accepted as a member of the 1968 entering class. Kinsey was the third woman admitted to the Institute in its twenty-nine year history.

Kinsey remembers her time in Appleton fondly. "It was a small community," she recalled. "I never felt like I was being given special treatment or that I was being discriminated against. We were all there to learn and we were all on the same basis. We all had nice housing, good facilities, and good professors. It has certainly been good for me to have had a broad education and I'm glad the Institute gave me the opportunity. I also think it was good for the Institute to begin to accept and welcome women students."

forty separate outfalls of ninety-one pulp and paper mills throughout the country to perform more than thirty thousand tests on approximately twenty-two hundred samples of paper industry effluent – all within a ten-month span. Three specific events led to the paper industry's panic and the demands upon the Institute. The first was the discovery of mercury contamination in the spring of 1970, the second was the federal government's reaction to Refuse Act 1899 in early 1971, and the third was the discovery of poly-chlorinated biphenyl (PCB) in foodstuffs beginning in 1971.

In early 1971, Institute researchers began a series of tests on the mercury levels in bodies of water upon which paper mills were located, in order to inform the public, the federal government, and the industrial consumers of paper, namely the food packaging industry.

The industry's demands became more frenzied a few months later as the Refuse Act gained national recognition. Two specific aspects of the Refuse Act were of particular significance to the paper industry; first, that a permit from the US Army Corps of Engineers was required before any industrial plant could dump refuse into waters, and second, that "waters" were defined as any "navigable waters or tributary thereof." This essentially included almost every running or still body of water in the country, with the possible exception of a small farmer's pond or a seasonal streamlet. The paper industry, which utilized close to one hundred tons of water for every ton of paper produced, became extremely concerned when executives realized that the permits required by the Refuse Act demanded a list of chemicals and various physical matter, including specific amounts, contained in effluent from every outfall of every plant. For Institute researchers, this translated into every sample requiring tests for up to fifty different parameters.

The Institute prepared by acquiring new and more sophisticated testing equipment, such as the Atomic Absorption Spectrometer, designed to detect mercury at levels of parts per billion. The Institute's staff was also realigned to deal with the testing and recording of results. Dr. Irwin Pearl was named coordinator of the program, while the majority of the testing was done under Dr. Dwight Easty's analytical chemistry group.

The original deadline for obtaining the necessary permits was July 1, 1971, which quickly proved impossible for Institute researchers. As Dr. Pearl commented in June 1971, "there isn't enough testing capacity in the country to take care of the situation." The testing process was complicated for a variety of reasons. Some of the samples required additives to ensure that they remained stable until testing time. Other samples were contaminated by extraneous material often found in the waters around pulp and paper mills, such as the noncellulose components of trees. Since test parameters assumed that chemicals were being checked in essentially clear water, such extraneous material often rendered the results invalid.

I think it was incredibly beneficial to take a field like chemistry and make it into something practical.

— **Jerry Sajbel,** entering class 1977

The family atmosphere…the warm association you have with the guys you went to school with and the professors at the school…the genuineness of it…cold winters, everybody's freezing, nobody had much money - but pulling together to make it a great experience. To me, that's what it was all about.

— **Gene Millard,** entering class 1970

The federal government permitted the deadline to be extended to October 1, 1971, for a portion of the test results. Members from other Institute research groups came to work in Dr. Easty's analytical group and staff members were put on three around-the-clock shifts seven day a week until the testing was completed. Not a single Institute client failed to meet the application deadline and receive the necessary permits.

The rising interest in environmental projects did nothing to detract attention from the Institute's other research projects, however. In 1970, the Institute nurseries were expanded to handle the production of improved hybrid aspens as part of the "Super Tree" program begun at the Institute in the 1960s. By 1972, more than one hundred and fifty thousand hybrids had been planted. Those planted at the program's inception now reaching an average height of forty feet and five inches in diameter. One thousand of these hybrids had survived at an altitude of two miles in Colorado in a climate ranging from forty degrees below zero to seventy degrees above.

In April of 1972, one of four test tube trees – the first to be produced by tissue culture – was planted in the National Arboretum in Washington D.C. in commemoration of the one hundredth anniversary of Arbor

Harry A. Posner, Jr.

Harry A. Posner, Jr., received his Bachelor of Science degree from the University of Southern California in 1958 and immediately began his career with Bell Telephone Laboratories as a member of the technical staff of Bell's White Sands Laboratory in New Mexico. At the same time, he began work on his Master of Science degree in electrical engineering at New Mexico State University.

In 1963, Posner was transferred to Bell's headquarters in New Jersey as supervisor of the Advanced Systems Design Group, where he was in charge of operation details and control systems for the Nike Zeus and Nike-X antiballistic missile programs. Posner remained in New Jersey until 1966, when he returned to New Mexico as director of the Physical Sciences Laboratory of New Mexico State University.

In 1968, Posner joined Stanford Research Institute as director of systems research and head of its Huntsville, Alabama, plant. He spent the next six years in a variety of positions within Stanford Research Institute, including assistant executive director of the engineering systems division in Menlo Park, California, and director of the Institute's Washington office where he was primarily responsible for relations with various government agencies. Prior to joining the IPC staff, Posner served Stanford Research Institute as the director of Omaha Riverfront Development Project, where he was responsible for the management of a six-county urban and rural planning and redevelopment program in the Omaha-Council Bluffs area.

In 1974, Posner was chosen to succeed John Strange as the Institute's fourth president. Posner led the Institute through its final Appleton era, turning the presidency over to Dr. Richard Matula in 1986. Posner was named President Emeritus and awarded an honorary Doctor of Science degree from Lawrence University.

Arnold E. Grummer

Originally from Latimer, Iowa, Arnold Grummer earned his bachelor of arts degree in English from the Iowa State Teacher's College at the University of Northern Iowa and his master of arts degree from Iowa State University in Speech and Creative Writing. He joined the staff of the Aid Association for Lutherans in 1957 as a member of the advertising department, a position he held until coming to the Institute in 1961.

At the Institute, Grummer held the title of Editor of General Publications. His responsibilities included the preparation and editing of non-technical institutional and academic publications as well as the preparation of news releases and other general public communications. In an interview with the Post-Crescent, Grummer once described his job as "basically that of communicating between a scientist and a non-scientist." In addition to his responsibilities as Editor, Grummer was also a faculty member in the General Studies department, where he worked with Institute students on both oral and written communications.

In 1961, Grummer began assisting the Institute's former Dean, Dr. Harry Lewis, in his duties as the Curator of the Dard Hunter Paper Museum. After Lewis's death in 1969, Grummer was named Curator, a position he held until his retirement from the Institute in 1975.

When asked about the general atmosphere of the Institute throughout the 1960s and 1970s in a 2005 interview, Grummer replied, "It was a complete intellectual environment. It was a think-tank rivaling anything existing anywhere on any subject."

Arild Miller

Dr. Arild J. Miller earned his Bachelor of Arts degree from Carleton College and his PhD in physical chemistry from Purdue University. After a stint with the Manhattan Project as a research chemist at Oak Ridge, Tennessee, during World War II, Miller joined the faculty of Carleton College's Chemistry Department.

Miller remained at Carleton College until 1960, spending his last several years there as the department head. In April of 1960, he joined the staff of the Institute as a member of the dean's office administrative group holding the title of associate director of admissions. He was named director of admissions in 1962 and was appointed assistant dean in 1964 in recognition of his numerous contributions to the Institute's academic program. He retired from the Institute in 1982 as the school's associate dean and was named Emeritus Professor of Chemistry.

I'll always remember Arild Miller as my introduction to the Institute.

— Al Jones, entering class 1962

I remember being very impressed with Arild Miller and the way he interviewed – he just presented himself really well, like he was in a class of his own.

— David Graves, entering class 1975

Day and in honor of J. Sterling Morton, who founded Arbor Day. Two more of the trees were planted on the Institute grounds, while the fourth was planted at the Wisconsin State Capitol Building in Madison.

The economic recession that ravaged the United States in the early 1970s had a profound impact on the Institute. By 1972, due to soaring inflation and a wave of mergers and consolidations, more than sixty member companies had disappeared from the Institute's ranks, leaving a huge dues deficit and eventually contributing to the fiscal crisis of 1974.

The crisis, blamed on the rapid acceleration of inflation and the lack of price increase recourse from educational institutions, was solved by the immediate contribution of hundreds of thousands of dollars from the Institute's member companies and, ultimately, the raising of member dues by twenty-five percent.

These events, as well as a series of sweeping administrative changes throughout the early 1970s, led to a radical restructuring of the Institute's administrative and academic policies. The research program underwent a shift from a program of individual and group contract research projects to a program of fewer but larger

I wouldn't trade it for anything. Getting my PhD opened up doors that would not have been there otherwise.

— **Rick Hartman,** entering class 1977

The Institute was a great way to start my adult life. It allowed me to have a great deal of fun while I was learning, then go out and have a great career. I feel I lucked into it.

— **Paula Stoppler,** entering class 1975

The whole concept of being a scientific generalist has been very useful to me. One of the things I really took away from it was the ability to think about things in a different kind of way, not just the straight and narrow.

— **Gail Carels,** entering class 1977

coordinated projects, reflecting the technological priorities of the pulp and paper industry. This shift was a result of the work of the Research Advisory Committee (RAC), a nine-member group created in 1971 to serve in a consultative capacity to the Institute's research, information, and service programs. By 1974, the RAC had discovered that the major areas of interest to the industry were research aimed at raw materials, reducing energy requirements, reducing capital intensiveness, board and paper properties, and specifications that are most economically related to end uses. Such findings led to the development of five major areas under which Institute research projects would fall: Raw Materials Research, Energy Research, Environmental Research, Capital Conservation Research, and Paper and Board Performance Properties Measurement. By 1976, the Institute was carrying sixteen major research projects that fell within one of the five categories.

The findings of the RAC and the subsequent shift in the Institute's research program in turn played a role in the restructuring of academic programs. An increase in student enrollment was a top priority during the first half of the 1970s, as the Institute, like so many other academic institutions, was still feeling the effect of the Vietnam War draft. Enrollment had climbed back up to its pre-war level of about seventy students by the beginning of the 1973-1974 academic year, but this still fell short of the Institute's goal of one hundred students, which would bring enrollment to capacity with the Institute's present facilities.

By 1979, the Institute's academic committee was forced to introduce a variety of curriculum changes designed to keep pace with the rapidly changing research interests of the industry. Among the changes were shifting the course in the physical chemistry of high polymers from an elective to a required course in

Hardev S. "Doug" Dugal

Originally from Dehra Dun, India, Dr. Hardev Dugal – known to Institutors as "Doug" – received his B.S. from Agra University, his A.H.B.T.I. in Chemical Technology from the Harcourt Butler Technological Institute in Kanpur, India, and his PhD from the Technishe Hochschule at Darmstadt, Germany.

Dugal's affiliation with the Institute began in November of 1964 when he received an appointment as a post-doctoral fellow in the physical chemistry section. He held this position until 1966 when he returned to India to conduct research.

Dugal returned to the Institute in 1969 and was appointed to the polymer chemistry group of the physical chemistry section. In 1975, based on his activity in both national and international environmental fields, Dugal was chosen to direct the Institute's new Environmental Sciences Division, a position he held for seven years. In 1982, Dugal was named director of the Information Services Division and head of the Institute's Department of Special Studies, where he remained until his retirement from the Institute in 1989.

the interdisciplinary core as well as the introduction of a new course designed to present a broad overview of the technical, political, social, and economic implications of environmental protection. Special emphasis was placed on the problems and consequences of such protection for the pulp and paper industry.

The 1970s marked yet another "changing of the guard" in the Institute's record books. In 1974, Harry Posner, Jr., formerly of the Stanford Research Institute, was chosen to succeed John Strange as the Institute's fourth President. Two years later, Dr. Harry Cullinan, the former chair of the chemical engineering department at the State University of New York at Buffalo, succeeded Dr. Roy Whitney as the Institute's Dean.

As the Institute's fiftieth anniversary approached and plans for the "IPC-50" celebration began to take shape, Bruce Smart, chairman of the academic policy committee, reminded industry executives that "our future needs for educated people, dedicated to paper, are as great or greater than they have ever been," thus setting the stage for the Institute's next fifty years.

John Strange

Born in Neenah, Wisconsin, in 1911, John Strange spent his entire professional career in a variety of positions at the Institute of Paper Chemistry. His first position, following his 1932 graduation from Lawrence College, was as administrative assistant in charge of paper codes. His administrative talents soon caught the attention of the Institute's top officers; indeed, Dr. Henry Wriston, president of Lawrence and the director of the Institute of Paper Chemistry, once said of him, "John Strange has become a skillful and an admirable administrative officer." Strange was selected as the Institute's secretary in 1936.

With the advent of World War II, Strange turned his attention to wartime duties. He played a key role in the creation of the War Products Development Section, Pulp and Paper Division, of the War Production Board, where, as the section's chief, he oversaw the development of some forty special paper products designed to meet wartime requirements. He resigned his position with the War Production Board in 1943 to return to his duties as the Institute's secretary. Following the resignation of Ralph Watts in late 1943, Strange combined his role as secretary with that of treasurer. In 1950, Strange's role at the Institute evolved into that of vice president and treasurer and, in 1955, following the retirement of Westbrook Steele, he was named the Institute's third President.

Strange's contributions to the Institute can hardly be underestimated. As the Institute's dean, Dr. Harry Lewis, once stated, "He [Strange] has a very real and constructive understanding of the purposes and objectives of the Institute and an unusual knowledge of the industry it serves." Strange's tenure as the Institute's president saw vast changes in both graduate education and in the United States paper industry. Strange handled these changes with his usual cool demeanor; under his leadership, the Institute thrived both academically and financially.

Strange retired from the Institute in 1974, turning the Institute's presidency over to Harry Posner. He was subsequently named President Emeritus and remained active in Institute affairs until his death in 1992.

Now that he is no longer with us, and it will not embarrass him to tears, I simply wish to say that I hold no man in the respect that I hold John G. Strange. His intellect, his erudition, and his person – I consider it an immense privilege to have had the opportunity to work with him. I consider him to have been the completion of my education.

— Arnold Grummer, IPC Editor of General Publications, 1960-1975

Mr. Harry [illegible]ner

It was a really interesting experience to be plowing along there, hoping that you can figure it all out - hoping that you can come up with an understanding, and a thesis, and data, something that people who have a range of views of the world will find compelling.

— **Alan Button,** entering class 1973

Mr. John Strange

What was really unique about the Institute was that it was a relatively a small group of people all studying a similar curriculum related to a particular industry. It was exciting because you knew that when you got out there were companies that were really interested in hiring you.

— **David Graves,** entering class 1975

Dr. Harry [illegible]nan

I really feel that the most valuable thing that I learned was how to solve problems. It's important to understand chemistry and science, but it's more important to know how to apply those things to problem solving. I discovered that once you got out into the real world of work that solving problems was really what it was all about. Teamwork was important there, just as it is important in the real world. I learned how to work well with others and how to engage people in a wide spectrum of activities.

— **Peter Balousek,** entering class 1974

Dr. Roy Whitney

The Institute's Night Before Christmas

'TWAS THE NIGHT before Christmas and in the IPC halls,
Not a creature was stirring, no one was washing the wall.
Beakers were hung in the racks with great care,
In hopes that the Nobel would soon be there.
The students were nestled all snug in their beds,
While visions of fellowships danced through their heads.
And Chuck in his last Opel and I in my cap,
Had just settled down for a long winter's nap.
When on top of the lab there arose such a clatter,
I sprang from my desk to see what was the matter.

Away to the stairwell I flew like a flash.
Tore open the door and fell on my…
The moon on the breast of the new fallen snow,
Gave a luster of midday to the objects below.
When what to my wondering eyes should appear,
But a new Buick Century and eight tiny reindeer.
With a tall, tanned driver so lively and quick,
I knew in a moment it couldn't be St. Nick!

More rapid than eagles, his coursers they came.
And he whistled and shouted and called them by name.
Now Wendall, now Doug, now Arild and Harry!
On Roy, on Earl, on John and Terry…
To the head of the Board, to a paper mill's call,
Dash away, dash away, dash away all!
As dry leaves before the wild hurricane fly,
When met with an obstacle mount to the sky,
So off to the airport the coursers they flew,
With cases of research and a proposal or two.

And then in a twinkling I heard in the halls,
The moans and groans an A-200 problem recalls.
As I drew in my head and was turning around,
My problem advisor came in with a bound.
He was dressed in lab white from his head to his foot,
And his clothes were all tarnished with precipitator soot.
A bundle of problems he had flung on his back,
And he looked like a peddler just opening his pack.
His eyes how they twinkled! His dimples how merry!
His cheeks were like roses, his nose like a cherry!
His droll little mouth was drawn up in a bow,
But the light in his eyes was cold as the snow.

He was chubby and cerebral, a right jolly old elf.
And I laughed when I saw him in spite of myself.
A wink of his eye and a twist of his head,
Soon gave me to know I had plenty to dread.
He spoke not a word but went straight to his work,
And passed out the problems, then turned with a jerk.
And laying a test tube aside of his nose,
And giving a nod, up the fume hood he rose.
He sprang to his desk, to his student gave a whistle,
And away they all flew like the down of a thistle.
But I heard him exclaim as they all ran in fright,
"Happy Christmas to all and to all a Good Night!"

The Institute Reporter
December 12, 1979

John Swanson

Born and raised in Sioux City, Iowa, John Swanson received his Bachelor of Science degree from Morningside College in 1940. After some graduate studies at Iowa State, Swanson joined the Institute staff in 1941 as a member of the physical chemistry group under Dr. George Sears. Much of Swanson's early research was focused on beater adhesives and he was soon known for his expertise on the uses of materials such as guar and locust bean gum.

In 1953, the physical chemistry group was renamed the colloid chemistry group with Swanson as its group leader. He was later named director of the Institute's division of natural materials and systems, a position he held until his retirement in 1982.

Swanson's career was filled with numerous awards and honors. In 1972, he was chosen as an Outstanding Educator of America as part of a program honoring distinguished men and women for their exceptional service, achievements and leadership in the field of education. He was presented with an honorary Doctor of Science degree from Morningside College in 1972 and an honorary Master of Science degree from Lawrence University in 1982. Considered an expert on paper surfaces, Swanson lectured at universities and research institutes throughout the United States, South America, and Europe.

John Swanson – he was absolutely the best. He was an outstanding professor.

—Robert Jones, entering class 1957

John Swanson was a very knowledgeable person. I think he was the most practical of the professors we had. He spent a lot of his time out working for member companies who had problems. Because of that, he could offer a great deal more practical knowledge about how things happened in paper mills.

—Carl Jentzen, entering class 1959

Swanson was one of my favorite professors, probably because even though he taught a lot of theory, he also tended to put things in a practical context.

—David Fary, entering class 1960

John Swanson was outstanding in the classroom.

—Jim Burke, entering class 1961

Swanson was a very effective lecturer.

—Harry Spiegelberg, entering class 1961

The thing I remember most about Swanson's class was his demonstration of the marangoni effect, which is when surface tension causes the alcohol in a martini glass to rise. We sat and watched martini glasses for a long time, waiting for the alcohol to go up and down.

—A[illegible] [illegible]hlen, entering class 1963

Even though he did not have a PhD, I would not exchange him for the knowledge of ten PhDs. He was a very good scientist and a very thorough scientist. He was a good teacher who conveyed the message extremely well to the students. What you learned from John were things you could apply later on in life while working in the industry. He wasn't theoretical - he was very practical. That was what we needed.

—A. Nurhan Becidyan, entering class 1969

The courses I took with John Swanson taught me how to be a scientist. I remember this one time that he asked me a question in class. I got it right and he asked me how I knew it was the right answer. I told him it was a gut reaction. I'll never forget what he told me – 'Gut reactions are not just guesses – they are learned reactions. You are actually bringing together all of your learning. It is not just a guess - finally your entire education is working for you.'

—G[illegible] Ring, entering class 1974

Outside the Classroom

Everyone was in the same boat - not a lot of money - so we did a lot of just getting together on Friday nights down at the Red Carpet Inn and making our own fun. Sometimes we'd have an official party, like for Halloween or Christmas, but most of the time it was just getting together to play cards or ping-pong and listen to records.

— **Gene Millard,** *entering class 1970*

We often spent a bunch of time unwinding in the Red Carpet Inn.

— **Paula Stoppler,** *entering class 1975*

We did a lot of fun things. I remember sledding – we took trays from the cafeteria and went sledding down the hill at Lawrence onto the river. We used to go cross-country skiing and snow-shoeing – a lot of winter sports. We often went to the movies and to bars - all the normal things.

— **Gail Carels,** *entering class 1977*

It was a great social scene. We always had a lot of fun.
— **Alan Button,** *entering class 1973*

We all lived in the student housing, so it was like our own little neighborhood.
— **Pete Balousek,** *entering class 1974*

CHAPTER 6

1980-1989

The 1980s played host to the shift between the industrial era that had characterized the United States since the Gilded Age and the new information age that would become the country's future. Silicon Valley emerged as the technology capital of the world as numerous and rapid advancements were made in sectors of technology ranging from medicine to electronics. Personal computers, video game systems, and even mobile phones targeted the modern consumer.

Like the rest of the country, the Institute of Paper Chemistry strove to keep pace with the rapid technological changes....

Energy was a hot topic in the United States throughout the 1970s and 1980s. Rising inflation, thriving commodity prices, the dominance of OPEC, the war between Iran and Iraq – all led to energy appearing, as Kenneth L. Fisher stated in an article in Forbes magazine, "unstoppable."

The Institute of Paper Chemistry followed suit. At the May 1980 Executives' Conference, Dr. Douglas Wahren, vice president of research, announced that the Institute's research program would be placing a much greater emphasis on energy via a three-year energy project subsidized by the United States Department of Energy. The project focused on four areas: a study designed to identify where energy goes in a paper mill to determine if energy flows could be altered or redesigned for greater process control and efficient utilization; the development of a means of energy accounting that would assign a value to energy in its various forms according to its ability to do work; a workable definition of energy leverage points; and the identification of specific projects to reduce energy consumption in the paper industry.

The early 1980s also saw the Institute striving to keep pace with the latest designs in computer technology. In 1980, a Burroughs B-6930 computer was ordered with funds from a $2.5 million fundraising drive begun in December of 1979. The new

computer, replacing the IBM computer used for the past twelve years, was delivered in 1981. It was more than ten times faster than the IBM, as well as having a vastly increased memory capacity. Designed to handle most of the Institute's scientific and administrative needs, the Burroughs B-6930 was able to

John M. Parker

Originally from Fairfield, Connecticut, John Parker received his Bachelor of Science degree from Central Connecticut State College and his Master of Education degree from Boston University. After staff positions at Deerfield Academy, Boston University, and Union College, Parker was named dean of student personnel services at Onondaga Community College in Syracuse, New York.

Parker came to Appleton in 1966 to succeed Edgar E. Dickey as the Institute's dean of students. He held this position until the Institute's move to Atlanta in 1989. Parker was named Emeritus Dean of Students upon his retirement.

When we were on the bush trip, John Parker made us introduce ourselves every time we stopped somewhere. It was a painful experience for me because I was so shy in those days, but, looking back, I guess it was one of the things that helped me get over my shyness.

— Gerry Ring, entering class 1974

John Parker always made sure to really welcome new students to the Institute. He wanted everyone to become familiar with the campus and with Appleton and everything.

— Chris Leutgeb, entering class 1985

What I remember the most about my introduction to the Institute was that John Parker, the dean of students, made it a point to pick me up from the airport himself. It really made me feel valued, like I was more than just a number to them.

— Jill Wallin, entering class 1988

communicate electronically with other data storage systems and computers in distant locations. Initially, the new computer was accessed through one of ten terminals scattered throughout the Institute; several more terminals were added by the end of the year.

By the mid-1980s, the Institute could no longer ignore the fact that the pulp and paper industry was in the midst of significant change. Foreign competition, the development of a world market, and a wave of mergers and consolidations all transformed the industry in numerous ways. These changes, in turn, affected the Institute, which depended on the industry for its survival. Static budgets, personnel reductions, and a need for major refurbishing of equipment beleaguered the administration and the board of trustees.

In February of 1986, Dr. Richard A. Matula, former Dean of Louisiana State University's College of Engineering, was named the fifth President of the Institute of Paper Chemistry, succeeding Harry A. Posner, who had held the title since 1974. Matula's first task was to initiate a strategic plan to set the stage for the future of the Institute by examining its needs and goals. By the summer of 1986, he had visited more than seventy pulp and paper industry executives to gather their thoughts and views on the direction of the industry and, subsequently, of the Institute. Utilizing Matula's findings as well as his own observations, a consultant was brought in to develop the strategic plan.

A task force was created in the spring of 1987, and by the following fall, was ready to present its conclusions to the Institute's board of trustees. A world-class pulp and paper institute that was capable of producing technologically excellent graduates who could also be team builders based upon cross-disciplinary knowledge was necessary, the task force concluded, in order for the United States pulp and paper industry to be competitive into the twenty-first century. Access to a large research university, the task force further advised, would be the key step in facilitating the plan.

The task force had already looked into several possible locations. The Institute wanted, according to an interview with Dr. Matula in 1989, "to be close to a university known for its accomplishments in technological research and to be close to a major city that could offer Institute employees a high quality of life." Further

It was a lot of fun – a good time in my life. The camaraderie among the students was excellent. It was a great education and it served you really well once you were out in the industry.

— **Steve Jones,** entering class 1984

criteria included proximity to other research centers and to an international airport with regular and reliable air service to different parts of the world. The Georgia Institute of Technology in Atlanta met – and exceeded – the criteria.

Georgia Tech and the State of Georgia were eager to cooperate with the Institute. The state offered more than $12 million in relocation funds, while the Georgia Board of Regents donated almost five acres of land for the construction of a new building and Georgia Tech offered to build a new parking lot adjacent to the new building. In addition, the Institute's member companies, noting that the new location was a hub of the pulp and paper industry with the potential to become an international pulp and paper center, agreed to pay double the amount of their annual dues for a period of two and a half years, a sum of more than $18 million.

On December 11, 1987, the Institute's Board of Trustees approved the plan to create an alliance with the Georgia Institute of Technology and relocate the Institute to Atlanta, stating that with the move, the Institute would be "establishing a close

Thomas M. Grace

Originally from Beaver Dam, Wisconsin, Thomas M. Grace received his Bachelor of Science degree in chemical engineering from the University of Wisconsin-Madison in 1960 and his PhD, also in chemical engineering, from the University of Minnesota in 1963. After receiving his doctorate, Grace spent two years as an aerospace technologist conducting experimental and analytical studies at NASA's Lewis Research Center in Cleveland, Ohio.

Grace joined the staff of the Institute of Paper Chemistry in December of 1965 as a member of the technology section. He was named chairman of the Institute's Department of Engineering in 1970, a position he held until 1974. Grace then served as a professor and a senior research associate until the Institute's move to Atlanta in 1989. He held an adjunct professorship at the Institute from 1989 to 1994, then a part-time position from 1995 to 1997. Upon his retirement, Grace was named Professor Emeritus.

Grace's research interests lay in the field of chemical recovery. In 1971, he began what proved to be groundbreaking research in the field by applying basic chemistry and heat transfer principles to the problem of black liquor fouling. The results of his research became the basis for improved design and operation of black liquor evaporators over next three decades.

Grace's career was filled with numerous awards and honors. While a professor at the Institute, he advised eleven doctoral candidates, all of whom received their PhD degrees. He taught five different academic courses in areas ranging from thermodynamics to the recovery of pulping chemicals. He received the AIChE Forest Products Division Award in 1984 and was named a TAPPI Fellow in 1987. In 2001, he was honored with the Gunnar Nicholson Gold Medal.

I had the good fortune to join Tom's research group at the Institute in 1975. I count the four years I spent there with Tom as the most important in shaping my career. Tom was a great mentor and a great teacher.

— Dr. William James Frederick, Jr., Director
Institute of Paper Science and Technology at Georgia Tech
2003 – present

Institute of Paper Science and Technology
formerly The Institute of Paper Chemistry
Appleton, Wisconsin

is pleased to announce its
relocation to

575 14th Street N.W.
Atlanta, GA 30318

Phone: (404) 853-9500
Fax: (404) 853-9510

Effective July 24, 1989

alliance with a major research university that has significant strengths in engineering and science, offering [Institute] students and professors access to computers and laboratory facilities that it could not afford in Appleton."

By the following spring, the decision was made to relocate the Institute in the summer of 1989. Each member of the Institute's two hundred and twenty-five member staff was invited to move with the Institute. Throughout the summer and fall of 1988, the Institute financed several short trips to Atlanta, allowing Institute employees and their families to see the area for themselves. Ultimately, more than one-third of the staff – the majority of them senior staff members – decided to relocate with the Institute, while the others opted for retirement or moving on to other jobs.

By the spring of 1989, the Institute had leased from the Georgia Tech Foundation a building on the corner of 14th Street and Hemphill Street to be used as a temporary facility until a permanent home could be constructed on the land donated by the Georgia Board of Regents at the corner of 10th Street and Hemphill Street on the western edge of the Georgia Tech campus. Renovation began in mid-April to ready the building for occupancy the following July.

In June, shortly after the Institute graduated its 1,302nd – and final – student from its Appleton campus, the school's accreditation was transferred from the North-Central Association of Colleges and Schools to the Southern Association of Colleges and Schools. June 30, 1989,

They trained me, they educated me, and they made me challenge myself. It was a very positive experience and I enjoyed it immensely. I would definitely say that I am the man I am today because I went there.

— **Russell Ross,** entering class 1988

It was a really positive experience for me – excellent preparation for my later training in chemical engineering.

— **Peter Ryan,** entering class 1983

Gwen Redding

Gwen Redding received her Bachelor of Science degree from Kansas State University and her Master of Science degree from Louisiana State University. From 1972 to 1975, she held two governor-appointed jobs in the Louisiana state government, first as the director of the Louisiana Bureau of Women and later as the deputy director of the Louisiana Division of Human Services. In 1975, Redding joined the staff of Louisiana State University as the director of Equal Opportunities Programs. In 1978, she transferred to the university's College of Engineering, where she served as the director of Engineering Services, assistant to the dean for administrative activities and the director of Minority Programs.

Redding came to the Institute in 1987 as a consultant. A year later, she was named vice president of administration and corporate secretary. As the Institute's chief administrative and financial officer, Redding was largely responsible for organizing the Institute's 1989 move to Atlanta, working closely with the moving company to facilitate the process. In Atlanta, Redding played a significant role in the interior design of the Paper Tricentennial Building, which was completed in 1993.

Redding's administrative duties at the Institute include direct supervision of the graphics and publications department, the library, the museum, building maintenance, and the abstractors. She left the Institute in 1996 to return to the private sector.

marked the last day of work for Institute employees not relocating to Atlanta, and on July 7, the first employees reported for work at the Industrial Research Facility to unpack the approximately one hundred truckloads of Institute furniture, equipment, and materials that made the nine hundred-mile trip from Appleton to Atlanta.

On August 1, 1989, the formal agreement between the Institute of Paper Science and Technology and the Georgia Institute of Technology was signed by IPST president Dr. Richard Matula and Georgia Tech president Dr. John Patrick Crecine. As Dr. Matula stated in an interview printed in the TAPPI Journal shortly after the signing, "With our new facilities and association with Georgia Tech, we will produce graduates who will help the industry be competitive into the twenty-first century." On September 20, the students of the Institute of Paper Science and Technology, sixteen of them first-years, began their first full day of classes in Atlanta.

The Institute was a family environment that prided itself on producing the best possible future leaders of the pulp and paper industry.

— **Mick Beck,** entering class 1985

Harry T. Cullinan, Jr.

Dr. Harry T. Cullinan, Jr., received his Bachelor of Science degree in chemical engineering from the University of Detroit in 1961 and his Master of Science and PhD degrees from the Carnegie Institute of Technology in 1963 and 1965, respectively. Before entering academia, Cullinan worked as a research engineer at Avon Cosmetics and Westinghouse Research and as a consultant to the Nuclear Materials and Equipment Company and the Western New York Nuclear Research Center in Buffalo.

In 1964, Cullinan joined the faculty of the State University of New York at Buffalo. He was named chair of the chemical engineering department in 1969, a position he held until his appointment to the Institute's staff.

In the spring of 1976, Cullinan was chosen to succeed Dr. Roy Whitney as the Institute's dean. In 1977, he was given the additional title of Vice President of Academic Affairs. Cullinan also served the Institute as a senior research associate, focusing much of his research on brownstock washing, stock refining, and recycled fiber. Cullinan left the Institute in 1987.

Earl W. Malcolm

After receiving his Bachelor of Science degree from Western Michigan University, Earl Malcolm entered the Institute of Paper Chemistry as a first-year student in 1959. He received his Master of Science degree in 1961 and his PhD in 1964.

Malcolm spent two years as a research chemist for Dow Chemical's James River Division and eleven years as a project manager in the product development department at Connecticut's C.H. Dexter and Sons before returning to the Institute as a senior research associate and director of the Institute's delignification division in 1976.

Malcolm advised more than fifty doctoral students during his twenty-four year tenure at the Institute. He held five patents and authored and coauthored numerous books and papers. In 1988, Malcolm received the Joseph K. Perkins Leadership and Service Award for TAPPI's Pulp Manufacture Division. He was named the Institute's associate Vice President of Research in 1999, a position he held until his retirement in 2000.

Dr. Malcolm had a wealth of knowledge and a great sense of humor.

— Brian Boyer, entering class 199

It was a sad and memorable time, wandering around the empty, silent buildings at the end of the first era of the 'Tute. There were laboratories, meeting rooms, and offices, which, once upon a time, were full of activity and excitement with experiments, lectures, PAC meetings, and seminars. It was a ghost structure, reminiscent of past glory and proud history. The laboratories were the places where many of the great inventions related to pulping and papermaking were made, the working environments of the brightest and most skillful people dedicated to our industry. The classrooms were places where the masters in paper chemistry, pulping, lignin, cellulose, hemicellulose, wood chemistry, paper physics, papermaking, and recovery taught. What a privilege to be their pupils!

— Nick Triantafillopoulos, entering class 1984

Richard Matula

Originally from Chicago, Richard Matula received his Bachelor of Science, Master of Science, and PhD degrees, all in mechanical engineering, from Purdue University. After completing his doctorate in 1964, Matula began his career in academia, teaching in a variety of undergraduate, graduate, and post-doctoral settings at the University of Michigan, the University of California, and Purdue University.

Matula's experience in academic administration began at Drexel University, where he served as the chairman of the mechanical engineering and mechanics department, director of the Environmental Studies Institute, and chairman of the thermal and fluid science group. He joined the faculty of Louisiana State University as the Dean of the College of Engineering in 1976.

Matula was chosen by the Institute's Board of Trustee to serve as President of the Institute of Paper Chemistry in late 1985 and assumed his duties in February of 1986. He left the Institute in 1995.

Dr. Matula was a visionary person with not only the perception but also the management ability to put necessary things in motion for accomplishing the vision. He often worked twelve and sixteen hour days and weekends in order to plan and outline the work necessary to attain change in institutions. Throughout his career, he was known as an administrator who could 'turn things around.'

— Gwen E. Redding, Vice President of Administration, Institute of Paper Science and Technology, 1987-1996

Dr. Matula was always smiling. I really enjoyed working with him.

— Connie Ford Adams, Administrative Assistant to the President, Institute of Paper Science and Technology, 1989 - present

Donald R. Dimmel

Donald R. Dimmel received his Bachelor of Science degree from the University of Minnesota in 1962 and his PhD degree in organic chemistry from Purdue University in [illegible]. After a postdoctoral position at New York's Cornell University, Dimmel spent four years as a senior research chemist at Hercules Inc., in Wilmington, Delaware. He was a member of the faculty at Marquette University when he accepted a position as a research associate and a professor of chemistry at the Institute of Paper Chemistry in [illegible].

Dimmel's research interests lay in the area of wood chemistry. He is considered an expert in the fundamental organic chemistry associated with pulping and bleaching reactions, including lignin and carbohydrate degradation reactions. While on the Institute's faculty, Dimmel served as an advisor to ten doctoral candidates and was honored with the Teacher of the Year Award twice. He authored or coauthored more than one hundred publications, including the book Lignins and Lignans, which was published in 2006. He retired from the Institute in 2002 as a professor, a senior fellow, and the leader of the process chemistry unit.

Dr. Dimmel was a character. I really enjoyed his sense of humor. He was an excellent pool player, too. I remember quite a few students being humbled by his skills at the pool table at the Red Carpet Inn.

— Tom Merchant, entering class 1987

Doc Dimmel was a great organic chemist.

— Brian Boyer, entering class 1991

I used to play racquetball with Dr. Dimmel at SAC. The problem was that I was 5'5" and Dr. Dimmel was over six feet tall. I was afraid that he was going to hit me when he was swinging his racquet with his long arms, so I would be way up against the wall of the court. Then he would make a nice shot to the opposite side of the court. I could never react fast enough to recover and he would most often win the point. It really was a lot of fun, even though I lost a lot!

— Karen Burnboom, entering class 1991

We spent a lot of time at the Red Carpet Inn, our own little bar and pool room in the basement of the resident hall. It was great place to relax and have a few beers after class.

— **Andrew Jones**, entering class 1985

physical game of tackle football in the snow.

— **Steve Jones**, entering class 1984

Outside the Classroom

Friday afternoons were always a great time at the Institute. The RCI usually stocked inexpensive beers, but on Friday afternoons, we would always buy a few cases of more expensive beer. It was a great chance to socialize with the other students and even some of the faculty members.

— **Chris Luettgen,** entering class 1985

Wisconsin was a great place. We were really a part of the town – you'd be surprised at how many Institute students married

The Bush Trip in the 1980s

I remember my time at the Institute as one of the few times in my life that I was solely focused on learning and not on worrying about what I was going to do afterwards. It was an outstanding experience and I wouldn't trade it for anything.

— **Pete Westervelt,** entering class 1980

My favorite memory of the bush trip is a discus-throwing competition that we had late one night. We used tree slices in lieu of an actual discus!

— **Andrew Jones,** entering class 1985

It was a great social occasion – a chance for us to get to know one another. Wisconsin is a fun place and we ended up going out a lot, which led to one really funny incident. One night, we stayed at this bar longer than anyone else. We were completely lost; we had no idea where Trees for Tomorrow was. We were wandering around the town trying to figure out how to get back when we came across the only building that still had its lights on – the police station. We went in and asked for directions. This one cop was kind enough to give us a ride back to the camp. I'm sure that made a great impression on our classmates and professors – us coming back in a police car!

— **Pete Westervelt,** entering class 1980

Since I had grown up in northern Wisconsin, I had been to most of those place a million times. But it was still an incredible start to the Institute experience. Of course, it was even more incredible for me because I met my future wife on that trip!

— **Russell Ross,** entering class 1988

It was a lot of tedious work, but it was also a lot of fun. I remember playing volleyball at the Eagle River camp and going into the town to see "A Fish Called Wanda," which had just come out.

— **Jill Wallin,** entering class 1988

I feel very fortunate to have had the opportunity to attend the Institute and learn everything that they were teaching there. Wherever it is located, I still feel the need to be a strong supporter of it, to give back to it in return for all it has given me.

— **Chris Luettgen,** entering class 1985

CHAPTER 7

1990–1999

If the key theme of the 1980s was rapid technological change, then the theme of the 1990s was the improvement and mainstreaming of those technolgies. The popularization of personal computers during the 1980s led to what is commonly know as the "dot-com boom" and the creation of the World Wide Web during the 1990s; an increased interest in the mysteries of outer space led to greater space exploration and a revolution in astronomy with the launching of the Hubble Telescope in 1990.

The Institute, too, faced what was perhaps the most significant change in its sixty-year history during the 1980s – a complete relocation. After settling in Atlanta during the summer of 1989, the Institute was faced with the challenge of making the most of its new facilities and educational program...

Students, faculty, and staff quickly settled into the routine of life at the new Institute of Paper Science and Technology. Classes were held and research was conducted – indeed, the pace was quite similar to that of the Appleton campus.

IN OCTOBER OF 1989, Charles E. "Chuck" Miller successfully defended his dissertation on the effects of polymer partitioning on fines retention, marking the first successful defense in the Institute of Paper Science and Technology's brief history. Five months later, on March 25, 1990, the Institute held its first commencement exercises in the Georgia Pacific Auditorium on Peachtree Street, awarding sixteen Master of Science degrees and six PhD degrees. A short six months after that, in September of 1990, a full class of thirty students began their coursework at the Institute.

James L. Ferris

Dr. James L. Ferris received his Bachelor of Science degree in chemical engineering from the University of Washington in 1966. After a stint at the Weyerhaeuser Company as a project engineer in the Everett Kraft Mill, he applied to graduate school at the Institute of Paper Chemistry and was accepted as a member of the entering class of 1967.

Ferris fondly remembers the familial atmosphere that was such a part of Institute life. "When you first arrived at the Institute," he recalls, "you were immediately welcomed into this huge, sprawling family. As soon as you got your keys from the dean and got settled, they would take you around and introduce you to everybody. Then they'd take your picture. As soon as they could, they'd distribute your picture to every professor, administrator, and secretary and they would all memorize your face and your name. So you'd be walking down the street or down the hall and people would call out 'Hey, Jim! How're you doing? All moved in? How do you like it so far?' It was simply amazing."

Ferris received his Master of Science degree from the Institute in 1969 and his PhD in 1974, spending the years 1970 to 1972 at the Naval Officer Candidate School in Newport, Rhode Island, and serving as a line officer on destroyers in the Western Pacific. He returned to Weyerhaeuser in 1974 as a bleaching scientist in the company's research division. A few years later, he became a global technical service manager, traveling around the world introducing new products and troubleshooting problems with existing products. Ferris spent the next fifteen years moving between a variety of positions in the company's research and business divisions. He was named vice president of research and development in 1988, a position he officially held until December 31, 1995.

In June of 1995, Ferris was asked to serve a temporary six-month term as the vice president of research at the Institute of Paper Science and Technology. He accepted the position and was instrumental in shaping the Institute's short-term strategic plan. Impressed with Ferris's experience and leadership skills, the Institute's Board of Trustees asked Ferris to become a candidate in the search for a new president. Initially hesitant, Ferris eventually entered his name. He was selected in late 1995.

Ferris's responsibilities as the sixth president of the Institute of Paper Science and Technology began on January 1, 1996. Working with Dr. Ronald Estridge, another Institute alum who had served as the Institute's interim president from June 1995 to December 1995, Ferris set as one of his immediate goals the re-creation of the familial atmosphere that he was so familiar with from his own time in Appleton. "I wanted to get the Institute back to some of its basic principles," he said. "When the students showed up to begin their graduate studies, I wanted everyone to know their names and faces, just like everyone at the Institute in Appleton had known mine."

Ferris retired from his position as the Institute's president in June of 2003. At the 2003 PIMA Annual International Management Conference, he was presented with the Lifetime Achievement Award in recognition for his years of dedicated service to the international pulp and paper industry.

I have nothing but the highest regard for Jim Ferris. He did an outstanding job managing things through a very difficult period and he was able to keep the Institute focused on both its educational responsibilities and its responsibilities to the industry.

— Paul Saueracker, Trustee,
Institute of Paper Science and Technology,
1997 – 2003

Dr. Ferris was a great president who always had a plan. And had it not been for him, I would have never read the book Who Moved My Cheese? *!*

— Connie Ford Adams
Administrative Assistant to the President,
Institute of Paper Science and Technology,
1989 - present

It quickly became apparent that the facilities at the Industrial Research Facility were ill-equipped for the number of students and faculty conducting research there. A new, much larger building was critical to the Institute's success.

Drawings and bid documents for the planned facility at the corner of 10th Street and Hemphill Street were finalized in February of 1990. A groundbreaking ceremony was held on May 10.

The Paper Tricentennial Building was completed in September of 1992. The 161,000-square-foot, five-story structure, built at a cost of $28.7 million, contained sixty-four laboratories, three classrooms, one hundred and seven private offices, six conference rooms, an executive conference room, a boardroom, a library, a large auditorium, a seminar room, and space for a museum.

The months of April and May of 1993 were a time of dedications and celebrations. On April 6, 1993, the library was dedicated in the name of Dr. William R. Haselton, who had made a significant gift for its development. Haselton, an Institute alumnus who had recently retired from a successful career in the pulp and paper industry, was honored with a plaque reading "This library is dedicated to William R. Haselton in recognition of his long association with the Institute of Paper Science and Technology as student, alumnus, trustee, and donor; and for his many professional and managerial achievements during the course of a distinguished career in the pulp and paper industry."

The people I met while I was a student at the Institute left a very profound impression on me. My classmates, my professors – I received what I feel is a world-class education.

— **Coray Harper**, entering class 1993

Two days later, the Institute held an enormous celebration in honor of the dedication of the Paper Tricentennial Building (PTB). Governor Zell Miller, former governor Joe Frank Harris, and Dr. H. Dean Propst, the chancellor of the Board of Regents of the University System of Georgia, attended the festivities. Finally, on May 6, 1993, a reception was held celebrating the completion of the American Museum of Papermaking.

The new state-of-the-art facilities in the PTB led to a massive reorganization of the Institute's research divisions. The Research Services Division was created to provide focus for research service activities within IPST and to enhance technical support for industry. Dr. Barry Crouse was named director of the division, which included the Institute's chemical analysis, microscopy and fiber analysis, optical properties, and paper analysis groups.

The Chemical and Biological Sciences Division was reorganized with Dr. Earl

Malcolm as its director to focus its efforts on improving raw materials and their conversion into pulp for papermaking and other applications. Chemical recovery, corrosion and materials engineering, forest biology, pulping and bleaching, and wood chemistry were all a part of the division. Finally, the Engineering and Paper Materials Division was created to focus on the manufacture, characterization, and performance of paper and related materials. Dr. Richard Ellis was named division director, supervising the container research, paper physics, papermaking, process simulation and control, and surface science groups.

Sweeping administrative changes were part of life at the Institute during the mid-1990s. In early 1995, Dr. Richard Matula, who had been the Institute's President since 1986, resigned from the position to return to the private sector. Shortly thereafter, Dr. Ronald Estridge, an Institute alum who had recently retired from a long and very successful career at the James River Corporation, was asked to serve as interim president for a six-month period until a new president could be chosen.

Estridge's time as the Institute's President, however brief, was the beginning of a new era at the Institute. As Dr. Robert Moran,

David Orloff

Originally from Philadelphia, Dr. David Orloff received his Bachelor of Science and Master of Science degrees, both in mechanical engineering, from Drexel University. He received his PhD degree in Applied Mechanics, also from Drexel, in 1974. His thesis research, supported by funding from the U.S. Air Force, involved elucidating the chemical kinetics of fluorocarbon and oxygen flames.

From 1974 through 1983, Orloff was a faculty member at the University of South Carolina, where he taught courses in thermodynamics, statistical thermodynamics, fluid mechanics, heat transfer, mass transfer, combustion, material science, and process design in the school's mechanical engineering department. In 1983, he joined the staff of the Kendall Company, first as the section head of the web forming and thermal bonding research section and later as the program manager of the non-wovens research section.

In 1988, Orloff became a faculty member at the Institute of Paper Science and Technology, teaching courses in thermodynamics, heat transfer, mass transfer, and papermaking processes. In 1992, he was named director of the engineering and paper materials division, a position he held until 1997.

Since 1988, Orloff has led the Institute's research efforts on the development of impulse drying technology, focusing on mathematical analysis and experimental investigation of the de-watering of paper via new pressing and drying concepts. His research resulted in eight United States patents and numerous associated foreign patents, as well as a series of pilot paper machine trials at the Beloit Corporation.

In 1999, Orloff was named the Institute's assistant vice president of research and technology transfer. Shortly thereafter, he became dean of research and technology transfer, responsible for the management of internally and externally funded research programs, interfacing with the member companies through the research advisory committee, working with federal agencies to identify new funding opportunities, and assuring the rapid transfer of technology to member companies.

Since the Institute's merger with Georgia Tech in 2003, Orloff has served as a professor of engineering in the George W. Woodruff School of Mechanical Engineering. His current research involves development of an understanding of pore expansion during flash evaporation in moist paper webs, which holds the potential to enhance the benefits of impulse drying.

The Institute is so unique - it's not your traditional academic institution.

— **Brian Boyer**, entering class 1991

Dr. Richard Matula

Dr. Ronald Estridge

Dr. James L. Ferris

entering class 1966 and past chairman of the Institute's alumni association, pointed out, "his [Estridge's] brief tenure was invaluable in bringing about a new and sharper focus on spending, research direction, and most importantly, a customer-focused approach to IPST's activities."

Estridge served as the Institute's President until January 1, 1996, when Dr. James L. Ferris, another Institute alum, took the reins. Ferris, who had worked for Weyerhaeuser in a variety of positions since 1974, most recently as Vice President of Research and Development, had come to the Institute during the summer of 1995 to serve a six-month term as the vice president of research. During the fall of 1995, Ferris was asked by the search committee to become a candidate for IPST president. He agreed, and was selected by the committee in late 1995.

Ferris went on to continue many of the programs initiated by Estridge, including the restructuring of senior faculty and research and technical staff and the revitalization of the Institute's employee council. Also important on Ferris's agenda was the creation of two vice presidential positions. Woody Rice, who had been with the Institute as a consultant since 1993 was chosen for Vice President of Development, while Dr. Gary Baum, a former Institute faculty member who had most recently served as the chairman of the Department of Wood and Paper Science at North Carolina State University, was named Vice President of Research and Academic Affairs. Gwen Redding, who had been with the Institute since 1988 as Vice President of Administration, left the Institute in 1995.

Within two years of Ferris's appointment as the Institute's President, the Institute had gained fifty-five new faculty and staff. His goal, which was attained in 1998, was a staff of two hundred and fifty.

The Institute's academic program experienced several major transitions in 1999. In August, Georgia Tech underwent what was commonly known as the "Semester Conversion," a program legislated by the Georgia Board of Regents to effectively end the quarter system. In order to maintain an identical academic calendar with Georgia Tech, IPST made the conversion as well. The Master's program, which had been a six-quarter program since 1992, was now four semesters. This led to a reorganization of some of the Institute's traditional courses, with several being merged to accommodate the longer semesters.

It was also during the fall of 1999 that the Management Sciences Certificate Program, which was begun in 1998, began to gain recognition. A joint program with Georgia Tech's Dupree College of Management, the certificate required two courses in Business Skills for Technology Managers and Leadership and Technology Management, both taught by IPST faculty, as well as a third course chosen from the Dupree College's elective menu on a subject such as negotiation and conflict resolution, international business environments, managing resources in the technology firm, or processes of technological innovation.

Gary A. Baum

Dr. Gary A. Baum received his Bachelor of Science degree in math and physics from Wisconsin State University and his Master of Science and PhD degrees in physics from Oklahoma State University. He joined the staff of the Institute of Paper Chemistry in 1969 as a research associate and an associate professor of physics. He was named group leader of the physical chemistry group in 1974 and was appointed director of the paper materials division in 1987.

Baum left the Institute in 1988 to join the James River Corporation as the director of Corporate R&D. In 1994, he left James River to become the chairman of the department of wood and paper science at North Carolina State University, a position he held until returning to the Institute as Vice President of Research and Academic Affairs in May of 1996.

Possessing excellent personal and technical reputations within the industry and academia, Baum succeeded in providing strong intellectual, technical, and administrative leadership in planning, coordinating, and maintaining the Institute's academic program and its research divisions. He retired from the Institute in June of 2002 to start his own company, Paper-Future Technologies LCC, which is devoted to applications of new techniques and uses for paper.

Gary Baum was a tough guy, but fair, very fair. He expected the best out of the students, and he usually got it.

— Rick Hartman, entering class 1977

Dr. Baum always amazed me with his high levels of professionalism and depth of knowledge. He was a fountain of information.

— Jerry Sajbel, entering class 1977

William R. Haselton

Originally from Glen Falls, New York, Dr. William R. Haselton became interested in engineering at an early age. After three years in the Navy, eight months of which he spent in the Philippines, he enrolled at Rensselaer Polytechnic Institute and received his Bachelor of Science degree in chemical engineering in the spring of 1949. He entered the Institute of Paper Chemistry the following fall.

Haselton remembers the challenges of being a student at the Institute. "The professors were very professional, very stern. Most had been teaching since before World War II, which took them back into a definite, distinct era in academia. They taught you well, but there wasn't any room for diversion."

Haselton still prides himself on the discipline he was generally able to maintain while he was a student at the Institute. "I was married, I had a child, I was living in a Quonset hut, and I had only a minimal income, so there was a lot of motivation there for me to finish up and get a real job," he recalls. Haselton's discipline ultimately paid off, for his tenure as an Institute student was filled with awards and honors. In 1950, he was the winner of the annual student award given by the Paper and Twine Association, selected by the Institute for exceptional accomplishment. In 1953, he was awarded the Westbrook Steele gold medal for outstanding thesis work for his research on gas absorption of wood and gas absorption techniques in studying paper pulp structure.

After graduating from the Institute in 1953, Haselton joined the staff of Rhinelander Paper Company as technical director. When Rhinelander was acquired by St. Regis Paper Company in 1956, Haselton went on to assume a variety of positions in the new company, including vice president and general manager of the forest products division and senior vice president for timberlands and forest products. In 1958, he was named executive vice president of operation, responsible for all operations of the Rhinelander plant with the exception of sales and corporate financial matters.

Haselton was elected Director of St. Regis Paper Company in 1972 and President in 1973. In 1979, he was elected Chief Executive Officer, and he was named Chairman of the Board in 1981. When St. Regis merged with Champion International Corporation in 1984, Haselton became Vice Chairman and a member of the board of board of directors.

Haselton's contributions to the United States pulp and paper industry have been numerous. He was named Papermaker of the Year in 1982 and Packaging Man of the Year in 1989. He served as president and chairman of the board of the National Forest Products Association and was a member of the executive committee of the American Paper Institute. Haselton was also a member of the IPST board of trustees from 1979 through 1988, serving on the executive committee and as chairman of the finance committee.

In 1992, Haselton contributed a substantial monetary gift to the Institute, stating that "several years ago I reached a decision that the Institute had made a major difference in my business career and my life in general. And for this reason, I became interested in showing my appreciation in some appropriate and tangible way." The William R. Haselton Library was dedicated on April 6, 1993.

Over the years, there's never been a doubt in my mind that choosing the opportunity for graduate study at the Institute was an extremely key element in what turned out to be an interesting and rather successful lifetime involvement with the forest products industry. I have to say it was one of the most stimulating and most rewarding and most interesting experiences of my life.

— Dr. William R. Haselton, April 3, 1993

Bill's ongoing generosity and commitment to IPST's future growth as the international center of technical excellence in the field of pulp and paper in one more milestone in the development of the Institute and a shining example of the ongoing support of IPST's alumni.

— Dr. James Ferris, President
Institute of Paper Science and Technology
1996-2003

Barry Crouse

Dr. Barry Crouse received his Bachelor of Science degree from Rutgers University in 1967, his Master of Science degree from the University of Maine in 1968, and his PhD from Duke University in 1972. He spent twenty-two years in the paper technology development division of Eastman Kodak, leaving in 1989 to join the staff of the Institute of Paper Science and Technology as a professor of chemistry and the director of the research services division.

Crouse left the Institute in late 1993 to establish his own consulting firm, but returned in July of 1995 to take the position of director of academic affairs. "I really wanted to be part of the team to help guide the Institute through this transition period," he said in 1995, referring to the sweeping administrative changes that were occurring at the time.

As the Institute's dean, Crouse focused on increasing the number of incoming students each year and on helping those students find positions in the industry following graduation. He was primarily responsible for the creation of a dual degree program with Georgia Tech and played a large role in integrating the Institute's academic program into Tech's. Dr. Crouse passed away in 2003.

Barry was the best sort of teacher, the kind who helped you set foot on the path of discovery to find out who you might be and what you might become. He would be uncomfortable to see you scarring your knees and scraping your knuckles, but he would not interfere with the growing process, beyond a few gentle but pointed words.

— Giselle Ow Yang, entering class 1994

Barry Crouse was one of my favorites. I always appreciated his guidance.

— Georgia Caruthers Malinsky, entering class 2000

Ronald Estridge

Originally from High Point, North Carolina, Dr. Ronald B. Estridge received his Bachelor of Science degree in chemical engineering from North Carolina State University in 1956 and his Master of Science and PhD degrees from the Institute of Paper Chemistry in 1959 and 1961, respectively. While at the Institute, Estridge was honored with the NPTA Award for Outstanding First Year Student and the Westbrook Steele Gold Medal for his thesis: "The Initial Retention of Fibers by Wire Grids."

After a stint in the United States Army stationed at Valley Station, Kentucky, Estridge joined the Albemarle Paper Manufacturing Company in Richmond, Virginia, as a senior development engineer. In 1974, he joined James River as the vice president of corporate development and executive vice president of the company's Massachusetts subsidiary. After serving the company in a variety of capacities, including group vice president of industrial specialties, executive vice president responsible for specialty and communications paper, and vice president and group executive of specialty paper, Estridge retired from James River as the senior vice president of technology in October of 1994.

In the spring of 1995, Estridge was chosen by the Institute's Board of Trustees to serve as the Institute's interim president following the resignation of Dr. Richard Matula, making him the first president who was also an Institute alumnus. Estridge was succeeded by another Institute alum, Dr. Jim Ferris, in January of 1996.

Everything I did was a true labor of love.

— Dr. Ronald B. Estridge, Appleton, Wisconsin, August 1996

Dr. Estridge was always extremely kind and thoughtful. I remember we had a United Way drawing and the prize was a roundtrip ticket anywhere in the continental United States. Dr. Estridge won the drawing! As he received his prize, we all heard someone in the back of the room shout 'give it to your secretary.' And he did! I couldn't believe it! I used the ticket to visit my parents in Arizona.

— Connie Ford Adams, Administrative Assistant to the President, Institute of Paper Science and Technology, 1989 - present

Richard L. Ellis

Dr. Richard Ellis received his Bachelor of Science and Master of Science degrees from Marshall University and completed his PhD at the University of Cincinnati in 1972. After fifteen years with Union Camp Corporation in a variety of technical positions, Ellis joined the staff of the Institute of Paper Science and Technology in September of 1989 as the director of the engineering and paper materials division, which focuses on the manufacture, characterization, and performance of paper and related materials. In addition to his duties as the division's director, Ellis was also a professor of engineering, with his research focused on paper physics, the dynamics of paper webs, and fluid mechanics.

In 1993, Ellis was named the Institute's vice president of research and academic affairs, making him IPST's chief scientific and academic officer. He left the Institute in 1995.

Dick Ellis was a very gregarious man, a very fun-loving guy. He had great rapport with the students.

— Brian Boyer, entering class 1991

Thomas J. McDonough

Dr. Thomas J. McDonough received his B.A. Sc., his M.A. Sc., and his PhD degrees from the University of Toronto, finishing in 1972. After several years in the paper and allied industries, he joined the staff of the Institute of Paper Chemistry in 1978 as a research associate and a professor of engineering.

McDonough was one of approximately eighty-five Institute faculty and staff members who relocated to Atlanta in 1989. Throughout his tenure, McDonough served as a professor, chairman of academic council, faculty chairman, associate vice president of research and academic affairs, and associate director of academic affairs. He directed numerous research projects in the science and technology of pulp manufacture, with emphasis on the technology, chemistry, kinetics, and environmental implications of pulp manufacture. McDonough served as advisor to sixteen doctoral candidates and published more than forty journal articles and book chapters throughout his career. He was named a TAPPI Fellow in 1995.

In 2000, based on his technical and academic experience, McDonough was named Center Director for the Center for Paper Business and Industry Studies (CPBIS). He retired from the Institute in 2002, although he continues to serve as the part-time Director Emeritus of CPBIS.

Tom McDonough likes to tell a story about how when people came to interview for faculty positions at the Institute, the hiring committees would always put them out in front of the Friday afternoon seminar crowd. Most of the graduate students were there. Tom says he vividly remembers me going after him about some of the ways he was presenting his information and analyzing all his answers!

— Alan Button, entering class 1973

Dr. McDonough was a fantastic teacher. He was really funny and he was easy to trust. He looked out for the students.

— William Allcock, entering class 1994

Tom McDonough had just joined the faculty as I was finishing up. He saw me through the last year or so of my thesis research. He was great – I had a lot of respect for him.

— Peter Balousek, entering class 1974

Dr. McDonough was usually pretty laid-back, but he was really intelligent and was good at explaining things. I always had lots of questions since I had never done any pulping before, but he made it really easy to learn.

— Sean Taylor, entering class 2000

Outside the Classroom

I spent a lot of time with the other students. We did a lot of whitewater rafting up in Tennessee and North Georgia and we'd go to Oktoberfest in Helen every year.

— **Brian Boyer,** entering class 1991

I spent more than twelve hours a day with other students. We went to classes together, studied together, worked on projects together, ate supper together, and studied some more. It wasn't all work, though – we played volleyball, softball and other intramural sports together as well.

— **Karen Turnboom,** entering class 1991

Most of the students formed study groups right away and that really built the camaraderie. We often took advantage of the things that Georgia Tech offered like bowling, the music room, and the gym.

— **Coray Harper,** entering class 1993

It wasn't all work. We usually spent our lunch hour playing cards. We played intramural soccer at Georgia Tech and hung out across the street at Billy Bob's a lot in the evenings. We used to go to this bar called Dirty Al's on Tuesday nights to play trivia. Usually we would win just enough to pay our bar tab.

— **William Allcock,** entering class 1994

The Bush Trip in the 1990s

It was a very rewarding experience. I truly feel that I wouldn't be in the position I am today had I not gone to the Institute. I learned a lot about the paper industry and its people. It helped me become confident in my abilities.

— **Karen Turnboom,** entering class 1991

We went down to the Florida panhandle and went around to several different pulp mills. We saw all kinds of large machinery, but since I was so new to the program, I couldn't distinguish one kind of machinery from another. It was a great chance to see what the industry was all about, though.

The bush trip was also a great time to establish good ties and strong bonds between the people that we would be studying with and working with for the next few years. It wasn't all just visiting pulp mills; we spent a lot of time just hanging out on the beach and hitting the beach bars at night.

— **Brian Boyer,** entering class 1991

The Institute was really good at teaching basic problem-solving - taking what you know, thinking your way through the problem, and asking the right kind of questions. It's the one thing I've used consistently throughout my life.

— **William Allcock**, entering class 1994

It was a great experience once you got used to the mosquitoes! It was sort of an initiation to the Institute, I guess – a good chance to meet your classmates and get to know people.

— **Coray Harper**, entering class 1993

The mills and the woodlands that we visited were interesting and a good introduction to the paper industry, but what I remember the most was body surfing in Pensacola!

— **William Allcock**, entering class 1994

I remember trips to several pulp and paper mills and visiting a forestry plantation where we learned how they geminated trees from seeds, grow them, and thin the crop. It was a great opportunity to learn more about the paper industry in the southern part of the country.

— **Karen Turnboom**, entering class 1991

CHAPTER 8

2000–present

The new millennium has seen an escalation of the numerous social issues that had plagued the latter half of the previous decade, especially the rise of terrorism and the brisk expansion of economic globalization. The widespread use of constantly improving communication devices such as mobile phones and the Internet have made the once-abstract concept of a global marketplace a reality.

Naturally, the creation of this global marketplace has had a profound impact on American industry, and the paper industry is no exception. Mergers, consolidations, and weak domestic and global economic conditions have led to a series of challenges unlike anything the American pulp and paper industry has ever faced before.

These challenges faced by the industry have, in turn, affected the Institute in numerous ways. After the Institute had lost nineteen of its fifty-three member companies to mergers and consolidations, it became more and more apparent that in order to survive, the Institute needed to consider an alliance of its own…

During the opening ceremony of the 2000 Executives' Conference, the theme of which was "Delivering Value to a Changing World," Institute president Dr. Jim Ferris spoke of the enormous changes that were transforming the structure of American industry. "The world is changing around us, and we must change as individuals, as companies, and as an industry faster than any of us would like," he told audience members. "And to make it more interesting, we really don't know how this transition will end up or how it will affect us as individuals, as companies, or as an industry."

THE INSTITUTE WAS NO STRANGER to change. Numerous developments in the paper industry, in education, and, indeed, in society in general throughout the previous seventy years had certainly given

William J. Frederick, Jr.

Originally from Bangor, Maine, Dr. William J. Frederick, Jr., received his Bachelor of Science, Master of Science, and PhD degrees, all in chemical engineering, from the University of Maine. After a series of non-academic research positions at General Motors, Battelle Memorial Institute, and United States Department of Agriculture's Forest Products Laboratory in Madison, Wisconsin, Frederick joined the faculty of the Institute of Paper Chemistry as a research associate in the materials engineering and processes division in the spring of 1976. Frederick stated that he took the position, one of several offered, because it was the most interesting and involved "the opportunity to do industrial research as well as teaching and working with graduate students."

In 1980, Frederick left the Institute to take a position as the group leader of the recovery technology applications division of the Weyerhaeuser Company in Tacoma, Washington. He left Weyerhaeuser in 1983 to rejoin academia as an associate professor of chemical engineering at Oregon State University. He became the department head in 1991.

In 1996, Frederick, who, in addition to his responsibilities at Oregon State had served in adjunct positions at both the Åbo Akademi University in Turku, Finland, and at the Institute of Paper Science and Technology, was approached by IPST President Dr. James Ferris about rejoining the Institute as a professor of chemical recovery engineering. "At that point," Frederick said, "I had basically rebuilt the entire chemical engineering department at Oregon State – I had hired half of the people in it. Things were going really well there and I wanted the opportunity for a new challenge. I went back to IPST because I thought that I'd have the opportunity to make a difference again." Frederick was named director of the Institute's chemical recovery and corrosion division in 1997, working in areas of chemical engineering, chemical reaction engineering, heat transfer, and electrolyte thermodynamics.

In August of 2001, Frederick was named the CEI Professor of Green Chemistry at Chalmers University of Technology in Gothenburg Sweden, where he remained until his appointment as the director of the Institute of Paper Science of Technology at Georgia Tech in September of 2003. "He [Frederick] brings an impressive set of credentials to the job of director. He's been a leading researcher and administrator, and he has strong ties to the Institute," said Dr. Ron Rousseau, the chair of Georgia Tech's School of Chemical and Biomolecular Engineering, upon Frederick's appointment. "Also important is that he is well-connected to the pulp and paper industry. He knows the strengths of IPST and he knows the opportunities for strengthening the delivery of research output to member companies."

In addition to his duties as the Institute's director, Frederick also holds the position of professor in the schools of Chemical and Biomolecular Engineering and Materials Science and Engineering at Georgia Tech. His favorite part of his job is without a doubt, working with the students. "I got into the education business and I stayed in the education business because I got such enjoyment out of working with young people and young minds," he says. "I enjoy watching those minds develop and be put to work. Students are usually so open to new ideas."

Jim Frederick and I played basketball together. We also liked to goof around when we were doing thesis work – that was generally how it was between Institute professors and students.

— Rick Harman, entering class 1977

Dr. Frederick is a great person, very fair and super to work with. I say 'work with' because he makes you feel like you work with him, not for him. And what a great family man!

— Connie Ford Adams,
Administrative assistant to the President,
Institute of Paper Science and Technology,
1989 - present

the school its share of transformations. The transformation that loomed on the Institute's horizon, however, was bigger and more disconcerting than any faced previously.

The Institute entered the new millennium still flying high from the honors and triumphs of the latter half of the previous decade. In May of 2000, the Institute was recognized as the fifth most innovative institution in the nation based on a survey from the Research Foundation of the State University of New York. With rankings based on the number of patents produced within an institution and the number of faculty at the institution, the Institute's fifth-place ranking came from its record of twenty-nine patents per one hundred faculty members for the 1998 research year. More than two hundred and fifty of the top research institutions in the country were included in the survey; the Institute placed ahead of MIT, Georgia Tech, Stanford, and Princeton.

The following fall, the Institute announced what was considered one of its highest achievements to date. On October 3, 2000, the Alfred P. Sloan Foundation declared that its board of trustees had approved the funding of its newest Center for Industry Studies. Named the Center for Paper Business and Industry Studies (CPBIS) and residing at the Institute and Georgia Tech, the new center set as its focus the business fundamentals impacting the global paper market with the purpose of generating research and producing experts for the paper industry. Five specific areas deemed of interest to the paper industry were chosen: globalization, commercialization, community interactions, workplace transformation, and enterprise effectiveness.

CPBIS was the sixteenth of the Sloan Foundation's Centers for Industry Studies, joining other centers at places such as

Harvard, MIT, Stanford, Wharton, and Columbia that are designed to create an academic community that understands industries and to encourage a direct approach to the companies and people of each industry for data collection and observation. The Sloan Foundation, a non-profit, philanthropic organization established in 1934 by Alfred P. Sloan, Jr., a longtime president and chairman of General Motors, granted CPBIS $2.04 million, which was then supplemented by $1 million in direct industry-based funding.

More than eighty-five Institute and Georgia Tech faculty members participated in the Sloan Foundation proposal. Dr. James McNutt, former president and CEO of Jaako Poyry Consulting, was named the Center's executive director, while Dr. Tom McDonough, a senior IPST faculty member, was named director. Dr. Gus Giebelhaus and Dr. Steve Vallas, both professors in Georgia Tech's School of History, Technology, and Society, and Dr. Vinod Singhal, a professor of operations management in Georgia Tech's Dupree College of Management, were named associate directors.

The development of the Center for Paper Business and Industry Studies represented a bond between the Institute of Paper Science and Technology and Georgia Tech

– a bond that was growing steadily as Institute faculty and students became more and more involved with, as IPST president Jim Ferris called them, "their Georgia Tech counterparts."

As the first year of the new millennium waned, it was obvious that the paper industry was going through a change unlike any it had ever experienced – and it was taking the Institute with it. Rapid consolidations and mergers within the industry resulted in IPST's loss of nineteen of its fifty-three member companies over a three-year period. With rumors of even more merger activity on the horizon, the Institute began to suffer the damaging effects of an enormous loss of revenue.

In the fall of 2001, the Institute's Board of Trustees asked the Institute to define what became known as "the Georgia Tech alternative" – that is, to examine the possibilities of an alliance between the two institutions. A team, sanctioned by the Chancellor of the University System of Georgia and working closely with the Institute's Alumni Association Executive Committee, was created to look into the possible benefits of an alliance, which, in addition to the financial benefits, included access to the broad array of new technology centers at Georgia Tech.

On September 24, 2002, after examining the prospects of alliance with several different colleges and universities, the Institute announced its decision to pursue a strategic alliance between IPST and Georgia Tech, effective July 1, 2003. Renamed the Institute of Paper Science and Technology at Georgia Tech, the Institute would become one of Georgia Tech's four large interdisciplinary research institutes, continuing to offer graduate degrees in paper science and engineering.

Dr. Tom McDunough

Dr. James McNutt

Dr. Jim Ferris

Sujit Banerjee

Dr. Sujit Banerjee received his B.Sc. degree in chemistry from the Indian Institute of Technology at Kharagpur in 1969 and his PhD in physical-organic chemistry from Montreal's Concordia University in 1974. After conducting postdoctoral research in physical-organic chemistry at McMaster University, Banerjee joined the staff of Syracuse Research Corporation in 1977 as a senior scientist. In 1983, he was named group leader of the Department of Nuclear Energy at the Brookhaven National Laboratory, a position he held until 1989.

An expert in environmental science with emphasis on the fate and transport of xenobiotics and aquatic technology, Banerjee joined the Institute staff in 1989 as the leader of the chemical analysis group. He holds several United States patents and has authored and co-authored numerous books, book chapters, and articles. In 1995, Banerjee, along with Dr. Terrance Conners of Mississippi State University, edited the groundbreaking Surface Analysis of Paper, which examines surface analysis techniques from a paper industry perspective and places heavy emphasis on applications. Banerjee was chosen by Institute students as the 2006 IPST Teacher of the Year.

In 2003, Banerjee joined as faculty of the School of Chemical and Biomolecular Engineering at Georgia Tech. His current research focuses on issues in the forest products industry with a strong emphasis on developing the fundamentals and taking them to commercial implementation. Banerjee's research group has set as its present goal the development of new techniques and novel materials for application in the pulp and paper industry.

The following December, Dr. Jim Ferris, who had been the Institute's President since 1996, announced his retirement effective June 30, 2003. An international search began immediately for someone to fill the position of Director of the new IPST at Georgia Tech.

In the spring of 2003, Dr. Ronald W. Rousseau, chairman of Georgia Tech's School of Chemical Engineering, was named the Institute's interim director, responsible for shaping the new organization during its transition period. Woody Rice, who had served as the Institute's Vice President of Development

since 1993, was elected President of the Institute of Paper Science and Technology in July of 2003, responsible for overseeing the smooth operation and transfer of various properties and assets from the Institute to Georgia Tech.

On July 17, 2003, it was announced that Dr. William "Jim" Frederick, Jr., the CEI Professor of Green Chemistry and Chemical Engineering at Chalmers University of Technology in Gothenburg, Sweden, had accepted appointment as the new Director of the Institute of Paper Science and Technology at Georgia Tech. He began his duties which, in addition to his responsibilities as the Institute's director, included a faculty position in Georgia Tech's School of Chemical and Biomolecular Engineering, on September 1.

The transition took approximately eighteen months. On December 14, 2004, Dr. Wayne Clough, Georgia Tech's President, and Woody Rice signed the final agreements integrating IPST into Georgia Tech. The integration was declared complete on December 31, 2004, and the Institute of Paper Science and Technology at Georgia Tech was left in the capable hands of Dr. Jim Frederick.

Georgia Tech is an outstanding school, and merging the research and academic operations of IPST with them is a very positive outcome in our search for a merger partner. A key element in our final decision was our desire to position the Institute where it can access the broadest set of advanced engineering skills needed by this industry to increase its global competitiveness. Georgia Tech has an outstanding science, computing, and engineering skill base that we can draw upon. The paper industry desperately needs innovative technologies, and we look forward to supporting their development through the new IPST at Georgia Tech.

— **A.D. "Pete" Correll,** Past Chairman, IPST Board of Trustees

The merger with Georgia Tech is a logical extension of our already close relationship that has developed since we moved to this campus in 1989. In becoming part of Georgia Tech, we simply take the final step to unite these two institutions. Our faculty and students are already heavily involved in activities with their Georgia Tech counterparts, and removing the organizational barrier will benefit us all as we continue to pursue our academic and research missions to serve our customers.

— **Dr. James Ferris,** President, Institute of Paper Science and Technology, 1996-2003

Elwood J. "Woody" Rice

Originally from Cornwall, Ontario, Canada, Woody Rice began his long career in the paper industry at a young age, working summers at Howard Smith Paper Mills. After attending Royal Military College of Canada at Kingston, Rice spent four years studying electrical engineering at Montreal's McGill University. He then went to work for Hercules Inc., a large chemical company based in Wilmington, Delaware. Rice spent most of his career in field and home office management in the paper technology group, which supplied paper chemicals to companies around the world. By the time Rice retired from Hercules in 1992, he had been promoted to group vice president of the paper technology group, responsible for its worldwide growth and operation.

Following his retirement, Rice was approached by IPST President Dr. Richard Matula and asked to aid the Institute in the development of some strategic marketing for the Institute membership. He soon became a fulltime consultant and, in 1993, was elected the Institute's new Vice President of Development, a position he held until the Institute's decision to merge with Georgia Tech in 2003. As the Vice President of Development, Rice was responsible for the development and maintenance of industry support, long-range government funding, and member company relations, all of which his long career and experience in the worldwide paper industry had prepared him for.

In July of 2003, Rice was elected President of the Institute of Paper Science and Technology, succeeding Dr. James Ferris. His role as President was to oversee the smooth operation and the transfer of various properties and assets from the Institute to Georgia Tech. In January of 2005, he handed the reins to Dr. Jim Frederick, who was chosen as the Director of the Institute of Paper Science and Technology at Georgia Tech.

Woody has many fond memories of his thirteen-year tenure at the Institute, including the privileging of awarding an honorary doctoral degree to George Mead II during the Institute's 74th Commencement exercises in 2004, which he calls one of the highlights of his career. He has been an avid supporter and life member of the American Museum of Papermaking, serving on the museum's Board of Trustees. Rice is also an active member of TAPPI and PIMA and has served on the boards of pulp and paper foundations at the University of Maine and North Carolina State University, as well as on the boards of various United States and foreign corporations.

David E. White

Dr. David E. White received his Bachelor of Science, Master of Science, and PhD degrees, all in chemical engineering, from the University of Pennsylvania. After completing his dissertation, "Diffusion-Limited Heterogeneous Catalytic Reactions on a Rotating Disk," in 1972, White spent four years as a research engineer at E.I. DuPont De Nemours and Company, where he was responsible for process development and scale-up.

White joined the staff of Union Camp Corporation as a research scientist in 1976. He spent the next twenty-three years with Union Camp in a variety of positions, conducting engineering research and design in papermaking, separations, reactor design, start-ups, and mill modelling for environmental compliance.

In 1999, White came to the Institute of Paper Science and Technology as the project leader for Project ROCIT (Redefining Our Core Industry Technologies), a venture designed to identify new technologies with potential to make a significant change in the cost competitiveness of the United States pulp and paper industry. In 2001, he was named director of technology transfer, where he was responsible for working with principle investigators to assure that research conducted at the Institute focused on the customer's needs.

White, whose areas of expertise include multi-phase pilot evaluation, computer modeling, papermaking, pulp bleaching, chemical and pulp/paper separations, and recycling, holds six United States patents and has been the recipient of several prestigious awards, including the American Chemical Society's Hero of Chemistry award in 1997. He was chosen as the Institute's Associate Director of Research in 2005. White's current research focuses on modified filler commercialization, future needs for digital printing, wood property effects on pulp yield, and forest products industry energy benchmarking.

Cyrus Aidun

Dr. Cyrus Aidun received his Bachelor of Science and Master of Science degrees from Rensselaer Polytechnic Institute in 1978 and 1980, respectively. He earned his PhD in mechanical engineering from Clarkson University in Potsdam, New York, in 1985.

Shortly after completing his doctoral research, Aidun was appointed a postdoctoral research associate in the chemical engineering department and a research fellow at the Mathematical Sciences Institute at Cornell University. He was then appointed a senior research consultant in computational fluid dynamics at the university's National Science Foundation's Supercomputer Center.

Aidun left Cornell in 1987, spending a year at Battelle Research Laboratories before joining the faculty of the Institute of Paper Science and Technology in 1988. As a professor of engineering, he taught courses in fluid mechanics, applied mathematics, fundamentals of liquid coating, thermodynamics, and heat transfer.

In 1992, Aidun was the recipient of the prestigious National Science Foundation Young Investigator Award, a highly competitive honor sought after by top faculty in major research universities. In 2000, he was responsible for the invention of a technology that essentially redesigned the flow mechanism in the fiber network-forming process. The result was a stronger paper made with less fiber and energy that, according to Woody Rice, the Institute's vice president of development, "will rank among the top advances in the papermaking industry over the last twenty-five years and should continue to be cutting-edge well into the twenty-first century."

In 2001, Aidun was chosen as the director of the Particulate and Multiphase Processes program at the National Science Foundation, where he managed the national funding of the program's research. Aidun was then named professor of fluid mechanics, paper science and engineering, and biotransport at Georgia Tech's George W. Woodruff School of Mechanical Engineering in 2003, where he had been an adjunct faculty member since 1989. His current projects continue his research in multiphase flow and have expanded into new multidisciplinary areas in biomedical research and nanoscale suspension flow with applications to value-added paper for cell culture in medical and pharmaceutical industries.

Cyrus Aidun was a very good teacher and an amazingly intelligent man.

— William Allcock, entering class 1994

Howard L. "Jeff" Empie, Jr.

Dr. Howard L. "Jeff" Empie received his Bachelor of Science degree from Syracuse University in 1964, his Master of Science degree from the Massachusetts Institute of Technology in 1965, and his PhD degree from the University of Minnesota in 1969. He worked in process research and development at the corporate research center of Hercules, Inc., from 1969 to 1978. In 1979, Empie joined International Paper as a senior research engineer in kraft chemical recovery at the company's corporate center.

Empie left International Paper in 1989 to join the staff of the Institute of Paper Science and Technology as a professor of engineering, teaching courses in reaction engineering, advanced chemical recovery, chemical engineering kinetics, mass and energy balances, and unit operations. Empie, who has advised numerous Masters' and PhD candidates during his tenure at the Institute, was chosen by Institute students for the 1994 Teacher of the Year award.

Empie joined the School of Chemical and Biomolecular Engineering at Georgia Tech in 2003 as a professor of chemical engineering. His current research focuses on reduction of white liquor deadload and purging chloride from the kraft liquor cycle, which involves using electrodialysis in a unique way that, in addition to solving a mill's chloride purge problem, should actually help mills show a profit.

Arthur J. Ragauskas

Born in Sudbury, Ontario, Canada, Dr. Arthur J. Ragauskas received his Honors B.Sc. degree in 1980 and his PhD in 1985, both in chemistry, from the University of Western Ontario. He spent two years as a National Science and Engineering Research Council of Canada (NSERC) Postdoctoral Fellow, studying carbohydrate chemistry at the University of Alberta, then another year in postdoctoral research at Colorado State University. In 1987, Ragauskas was named a National Research Council research associate, working in the field of immunochemistry and concentrating on the synthesis of salmonella trisaccharide epitopes.

Ragauskas joined the Institute staff in 1989 as a professor of wood chemistry, conducting research on the fundamental principles involved in the photoyellowing of mechanical pulp.

Ragauskas has supervised eighteen PhD students, sixteen Masters' students, and seventeen postdoctoral research fellows throughout his tenure at the Institute, several of whom have received national and international awards. He has published more than seventy journal articles and presented his research at numerous national and international conferences. In 1999, Ragauskas was chosen by Institute students to receive the Teacher of the Year Award, and in 2004, he was honored with the prestigious TAPPI Fellow Award, given in recognition of meritorious service to TAPPI and to the pulp and paper industry.

In 2003, Ragauskas became a member of the faculty of Georgia Tech's School of Chemistry and Biochemistry. His current research is in the field of carbohydrate chemistry, with the primary objective to harness the possibilities inherent in the chemistry of cellulose, the most abundant of Earth's organic chemicals.

Dr. Ragauskas was quite a character. We always laughed a lot.

— Brian Boyer, entering class 1991

The professor I remember the most is Dr. Art Ragauskas. He was my advisor and he left a huge impression on me. He taught me how to schedule my classes, which was really helpful while I was in school. He also provided me with a perspective of the industry in terms of keeping your eyes a little bit ahead and keeping your ear to the ground, so to speak, in order to keep your career fresh – how important it was not to get stale.

— Coray Harper, entering class 1993

The Robert C. Williams
Paper Museum
at Georgia Tech

I think the American Museum of Papermaking is the soul of the Institute. We see the history of not only IPST, but the history of the paper industry reflected in the Museum. We see the beauty that paper can bring to the world. It is not just an attachment; it is a very important part of what we do. It's very important for those of us who spend most of our time in the science and engineering worlds – it brings the artistic world to us in a very nice way. I'm very glad we have the museum. I've often told people that I think that's the best part of the deal that Georgia Tech got when they agreed to take on IPST.

—Dr. William J. Frederick, Director,
Institute of Paper Science and
Technology at Georgia Tech,
2003 – present

The idea of a museum dedicated exclusively to paper and paper-related artifacts is almost as old as the Institute itself. A wasp's nest representing papermaking in nature was accepted as a donation by the Institute's Technical Director, Dr. Otto Kress, in 1934; two years later, Mr. Westbrook Steele, the Institute's Executive Secretary, accepted a donation of 1871 commercially made sulfite-pulp paper. More items, including microscopes, fibers, and testing equipment, were added to the collection over time and found a home in a glass display case near one of the Institute laboratories.

In 1939, at the urging of Massachusetts Institute of Technology President Dr. Karl Taylor Compton, Dard Hunter, a world-renowned expert on the history of paper and papermaking, established a paper museum at MIT in Cambridge, Massachusetts. His goal, which remains part of the museum's mission statement to this day, was to "stimulate interest in the ancient craft of papermaking and promote understanding of present-day paper…" Artifacts representing papermaking activities in countries around the world were artfully arranged in glass display cases by Hunter himself.

By 1950, however, Hunter had become disillusioned with slow trickle of visitors the museum was receiving. The retirement of Dr. Compton, a long-time advocate of Hunter and his collection, did not help matters. By 1951, the collection had been moved to a much smaller facility in the school's Hayden Library and the Dard Hunter Paper Museum had become a seldom-used and somewhat obscure research center.

To collect, preserve,
increase and disseminate knowledge
about papermaking –
past, present and future.

I remember being able to go down to the basement and visit the museum while it was in Appleton. It was great there, but there is really no comparison to how wonderful it's become here in Atlanta. It's a fantastic resource for the Institute, for the public, and for the industry. Cindy [Bowden] has done a wonderful job of capitalizing on all of the museum's assets. She's turned the collection into what is now truly a global resource.

— **Dr. James Ferris,** President, Institute of Paper Science and Technology, 1996-2003

As far as I am concerned, the Institute advanced a long way in the building up of its national and international stature through the acquisition of the Dard Hunter Museum. Here it occupies an area of approximately four thousand square feet in quarters largely designed for it. This collection is the world's finest…it should draw many people to the Institute.

— **Dr. Harry F. Lewis,** Dean, Institute of Paper Chemistry, 1930-1956

It wasn't long before the Institute learned of the floundering situation at MIT. Institute Vice President and Treasurer John Strange began correspondence with Dard Hunter shortly thereafter and, by 1954, the Institute had offered to reestablish the Dard Hunter Paper Museum in Appleton. The arrangement was finalized on July 30, 1954, with Hunter himself named curator of the collection, which was displayed in a bomb-proof room in the basement of the General Activities Building. The museum officially opened to the public in early 1955. Hunter was quite pleased with the arrangements in Appleton; as he told a Peruvian acquaintance, Victor von Hagen, in a letter dated February 17, 1960, "the paper museum is now serving a more useful purpose than at MIT, as in Appleton it is seen by many more interested people than formerly."

Dard Hunter was the museum's curator until his death in 1966, at which time the Institute's former Dean of Students, Dr. Harry Lewis, who had retired from his position in 1956 and had been assisting Hunter with the collection for more than ten years, took over for a two-year period.

The museum grew by leaps and bounds over the next twenty years as the collection was increasingly enlarged by donations of various paper and paper-related artifacts. As Doug Stone, a museum advisory board member who also assisted with the first major attempt to catalogue the collection in 1980, pointed out, "gradually, others in the local paper community became more aware of who Dard Hunter was and the extensiveness of his work. This inspired them to look at their own collections and periodically donate artifacts, paper specimens and books."

When the Institute made its nine hundred-mile move to Atlanta in 1989, the Dard Hunter Collection was packed away as well – a major undertaking that required the services of a variety of museum and preservation specialists. Once in Atlanta, the facilities of the newly renamed Institute of Paper Science and Technology proved too limited to accommodate the collection. It remained in boxes until construction of the Paper Tricentennial Building, the Institute's

Dard Hunter was very well known and respected throughout the papermaking community, much of which was centered in Wisconsin's Fox River Valley. Hunter was an artist of paper history, someone they learned a lot from, who traveled widely and understood the history of different papermaking cultures. Needless to say, everyone in Appleton and the surrounding areas were thrilled to learn that the Institute was to acquire Dard's collection!

— **Doug Stone,** Advisory Board Member, 1995 - present

Dard Hunter

Originally from Steubanville, Ohio, William Joseph "Dard" Hunter was absorbed into a world of paper and printmaking at a young age as he routinely assisted his father in the family's newspaper business. After a short period as a staff artist for his father's Chillicothe, Ohio, paper, Hunter set out to see the country as an assistant to his brother, Philip, an accomplished magician. It was during one of these tours that Hunter became fascinated with the American Arts and Crafts Movement.

In the early summer of 1904, Hunter applied for a summer internship with Elbert Hubbard's Roycrofters in East Aurora, New York. Hubbard, a leading proponent of the Arts and Crafts Movement, at first denied Hunter's application, but Hunter, ever persistent, simply traveled to East Aurora anyway and was eventually hired and began to develop his artistic talents in a variety of media, including graphic design, stained glass, and metalwork.

Hunter became captivated by early European papermaking and printing during a 1911 trip to London and was overcome by a desire to learn more about the centuries-old art of hand papermaking. Upon his return to America, Hunter and his wife, Edith, moved to Marlborough, New York, to build a paper mill on Jew's Creek with the goal of producing handmade paper for American artists and printers who had, until then, relied upon European papermakers to fulfill their handmade paper needs. Relying exclusively upon seventeenth century hand papermaking techniques, such as depending entirely upon a water wheel to power the mill, Hunter continued to practice the creed he had learned at the hands of Elbert Hubbard and the Roycrofters - that a hand-crafted object was inherently more desirable, beautiful, serviceable, and worthy of human endeavor than anything made by a machine.

In 1919, Hunter returned to Chillicothe, where he purchased the Mountain House, which was built in the early 1850s. He immediately reconstructed one of the wings to form a letterpress printing studio, which he named Mountain House Press.

Dard Hunter devoted the rest of his life to researching, collecting, writing, and publishing the world's history of hand papermaking and printing, crisscrossing the globe dozens of times to some of the most remote places on earth to study the world's various papermaking cultures. He eventually authored twenty books on paper and papermaking, eight of which were printed by hand. The artifacts and paper samples he collected on his journeys form the bulk of the Dard Hunter Collection, originally housed at the Massachusetts Institute of Technology and now featured in the Robert C. Williams Paper Museum at Georgia Tech in Atlanta.

permanent home, was completed in late 1992. A reception celebrating the official grand opening of the American Museum of Papermaking was held on May 6, 1993.

In the spring of 1996, the museum, which was receiving more than ten thousand visitors annually, was officially renamed the Robert C. Williams American Museum of Papermaking, the result of a significant endowment gift from the James River Corporation in recognition of its cofounder Robert C. Williams, who had retired as the company's chairman, president, and CEO in 1995. The endowment played a large role in the expansion of the museum's new traveling exhibition program, which was met with remarkable accolades.

The many rare artifacts that traveled from Appleton are now part of a permanent collection in Atlanta tracing the fascinating history of papermaking from its earliest antecedents to the advanced technology

of today – from ancient Summerian cuneiform tablets and Egyptian papyrus scrolls to the exquisite chiaroscuro watermarked papers still made in Fabriano, a well-known papermaking center in the Marches region of Italy. In 2005, the George W. Mead Education Center, created with generous funds from the Mead-Witter Foundation, doubled public space capacity, which has enabled the museum to increase educational and public programming while also allowing greater public access to artifacts from the permanent collection as well as to the museum's impressive rare book collection.

The museum's critically acclaimed changing exhibition gallery showcases various interpretations and uses of paper as a viable and diverse medium by hosting exhibitions by contemporary artists working in the allied fields of paper, printmaking, and book arts.

Through the offerings of the gallery, museum visitors are not only able to view artistic applications of many of the processes described by the interpretive display of the expressive uses of paper, but also to connect to paper as a living art, not just an ancient craft transformed through technology to purely utilitarian uses. Recent exhibitions have included brilliant cast pulp paintings, an outstanding selection of the finest work in handmade paper created by art students

The Dard Hunter Paper Museum shows the antiquity and the universality of paper. By a sweep of the eye, a visitor may scan centuries and in several steps he can cross whole continents engaged in this ancient craft. The museum tells us more than the story of paper and its evolution. Paper has been a part of man's culture. To know the story of paper is to see man keeping records, engaging in commerce, creating literature, responding to beauty and worshipping a deity.

— **Dr. Edwin W. Schoenberger,** Dean, Institute of Paper Chemistry, 1956-1960 from throughout the United States in the museum's tri-annual collegiate handmade paper art show, an exhibition of exquisite sculptural book forms, a display of richly embellished collages and assemblages incorporating handmade paper and found objects, marvelous colored pulp paintings with environmental images and themes, handmade paper quilts, and a selection of fine handmade books and book designs created by educators throughout the Southeast. The gallery also features scholarly exhibits, which illustrate various cultural traditions of papermaking and an examination of papermaking from a historical perspective. A variety of artist-led workshops and lectures accompany each

exhibition. In hosting these exhibitions, the museum provides the public with a broader experience while giving artists the opportunity to display their art in a space that connects their work historically, physically, and conceptually.

Outreach is a vital part of the museum's mission. In addition to the thousands of school children that visit the museum for guided tours each year, more than one million visitors have experienced the world of paper and papermaking through the museum's two traveling interactive exhibitions. In the first interactive exhibit, called PaperQuest, children are able to crawl through five sections of a paper machine, play computer games in a paper mill control room, play games about where paper can be found in the home, and see artifacts in the global history area. They are given the opportunity to make a piece of paper from recycled blue jean material at the papermaking station. The second interactive exhibit, A Child's View of Papermaking, allows children to climb down from the top floor of the two-story exhibit through the drying cylinders of a paper machine while learning about the different professions in papermaking.

Another traveling exhibition, The World of Paper, was seen by more than eleven million visitors during its eighteen-month display at Atlanta's Hartsfield-Jackson International Airport. Featuring artifacts from the museum's permanent collection as well as a series of handmade paper artwork, the exhibition then traveled to the Museo de Bellas Artes in Santiago, Chile, where it was met with the largest opening reception in the history of the museum. An upcoming exhibition partners the Robert C. Williams Paper Museum at Georgia Tech with the Museo Della Carta e Filligrana Italy with the intention of creating an international virtual exhibition to focus on the little understood but significant contributions of paper to the evolution of the European Renaissance. The new exhibit will explore a nearly unexamined view of paper's arrival in Italy and its spread through Europe during a time when scholarship flourished and books, the printed word, and literacy extended their influence.

The Robert C. Williams Paper Museum at Georgia Tech fulfills a vital role as a communicator and a vehicle for the preservation and study of paper-making's important past and continuing contribution to our culture. By allowing visitors to examine paper from its origins to its contemporary interpretations, the museum is a monument to the rich history, the global present, and the very bright future of a product to which we are all connected.